Calvert Vaux

Villas and Cottages. A Series of Designs

Calvert Vaux

Villas and Cottages. A Series of Designs

ISBN/EAN: 9783337188108

Printed in Europe, USA, Canada, Australia, Japan

Cover: Foto ©berggeist007 / pixelio.de

More available books at **www.hansebooks.com**

n

FAMILY COTTAGE IN THE MOUNTAINS.

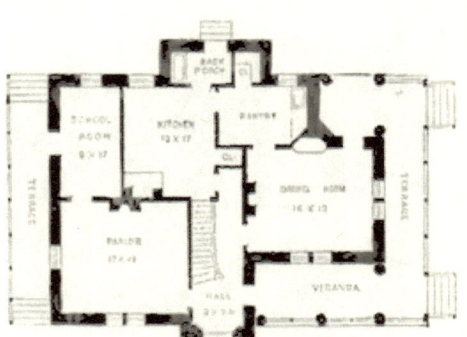

PLAN OF PRINCIPAL FLOOR.

VILLAS AND COTTAGES.

A Series of Designs.

BY CALVERT VAUX,
ARCHITECT,
OF NEW YORK.

An entirely New Edition.

ILLUSTRATED BY 370 ENGRAVINGS.

NEW YORK:—HARPER AND BROTHERS.
LONDON:—SAMPSON LOW, SON, AND MARSTON,
14, LUDGATE HILL,
ENGLISH, AMERICAN, AND COLONIAL BOOKSELLERS AND PUBLISHERS.
1864.

Inscribed,

WITH SINCERE RESPECT.

TO

CAROLINE E. DOWNING,

AND

TO THE MEMORY OF HER HUSBAND,

ANDREW J. DOWNING.

PREFACE.

EVERY American who is in the habit of traveling, which is almost equivalent to saying every American, must have noticed the inexhaustible demand for rural residences that is perceptible in every part of these Northern States. Nothing like it has ever yet occurred in the world's history; and although hard times undoubtedly occur in America, as well as elsewhere, at occasional intervals, it would seem that the profits which are missed by one man, contrive, somehow, to slide into the pockets of other more successful operators; for the carpenters and masons appear to be always getting a full percentage of the floating capital, and the ball is kept merrily rolling under all changes of individual circumstances.

Such being the fact, whatever may be its philosophy, it seems evident that the season must come when the importance of the whole subject of domestic architecture will be fairly and fully recognized. It can not be possible that the energetic vitality which pervades this branch of home manufacture will, for any great length of time, remain satisfied to expend its intensity on meagre, monotonous, unartistic buildings, or that it will continue to pay out millions of dollars every year without perceiving the propriety of getting, habitually, something worth having for the moncy. In an intelligent age and country like this, ugly buildings should be the exception: not, as hitherto, the almost invariable rule.

The accompanying designs have been prepared within the last few years to respond to the varied requirements of different parties who have asked for them, and it is conceived, therefore, that they may possibly represent, to some useful extent, to those who are about to build in the country, the accommodations and arrangements for convenience that appertain to such buildings. They are not brought before the public as model designs, to lessen the necessity for the exercise of individual taste, but, as far as possible, to increase its activity. Such books are needed as stepping-stones; for no general popular progress can be made in any art without ample and cheap opportunities for comparison and criticism; and the chief value of illustrated works on such topics as domestic architecture must always lie in the fact that they are calculated to rouse into active life the dormant capacity for individual preference, which all possess more or less, and which is absolutely necessary for a just artistic opinion on any subject. It is for this reason, and with the hope of being more generally intelligible and popularly useful, that the engravings are arranged, in the present volume, in a condensed, regular manner, so that they may be examined with little trouble and with but slight reference to the descriptions; the eye thus being enabled to glance from one to the other briefly and easily.

In this collection of studies there are many marked "D. and V." that have a special interest as the latest over which the genial influence of the lamented Downing was exercised. Several of the plans were in progress when the tidings of his sudden and shocking death were mournfully received by his family and friends, and almost as mournfully by thousands, who, knowing him only through his books, still felt that he was to them a dear and intimate companion. Mr. Downing was on his way to Newport, to superintend the execution of Mr. Parish's villa, on the day when the loss of the *Henry Clay* in an instant struck out his name from the roll of living men, and thereby inflicted an irreparable injury on his country; for Andrew Jackson Downing was not only one of the most energetic and unprejudiced artists that have yet appeared in America, but his views and aspirations were so liberal and pure that his artistic perceptions were chiefly valued by him as handmaids to his higher and diviner views of life and beauty. It is for this reason that his loss is so severely felt; for his character being moulded on this large scale, and his capacity to appreciate whatever is beautiful in nature or art being proportionately great, he had both the will and the power to exercise a marked influence for good over the taste of his countrymen. He readily saw that the contempt of art which the early Puritans had shown with the best intentions, and which in their age of imperfect toleration was entirely intelligible, and perhaps necessary, exists now only as a chronic, unmeaning prejudice; he also perceived that the proper time had arrived for the exercise of a better state of feeling, and for a general popular advance in taste, and he addressed himself to the furtherance of this good work with quiet enthusiasm. He used every effort to break down the foolish barrier that ignorance had set between the artist and the moralist, and strove to make manifest in all his works the glorious truth that the really "beautiful" and the really "good" are one. This conviction is indeed the key-note to all his teachings. "*Il bello e il buono,*" was the motto engraved on his seal and on his life; and the everyday increasing improvement that is now visible in the popular taste, so far as regards the subjects on which he wrote, and which may be directly traced to his books, is ample evidence that his modest, earnest words must have sunk gently and convincingly into the hearts of many worthy readers. He was fortunately not a man of promise only, but of rich performance; and although cut off in the very prime of a hearty, active, ever-expanding life, he had already lived and labored to such good purpose that he can scarcely be said to have left his work unfinished. He has set his mark fairly and broadly on the spirit of his age, and it is to be hoped that the love for grace and beauty that he so vigorously aroused in America will in future be always advancing.*

A few of the studies submitted have been made especially for this work, to illustrate particular views; but the greater number are reduced from working plans of designs that have been either executed or prepared for execution, and whenever practicable, the particulars of contract or expenditure are supplied. Some of the designs, it will be observed, are marked "V. and W.;"

* Some time after the loss of the *Henry Clay* a private subscription was raised for the purpose of erecting, in the grounds attached to the Smithsonian Institute at Washington, some fitting memorial of Mr. Downing, who was engaged by the government, at the time of his death, in carrying into execution a comprehensive plan for landscape gardening that included the Smithsonian grounds, and also the whole of the public park proposed to connect the President's house with the Capitol. The design ultimately determined on for this memorial, which is now being erected at Washington, on the site appropriated for the purpose, is illustrated by the vignette on page xii. It is simply a large, white marble vase, carefully modeled from a chaste but highly enriched antique example, and mounted on an appropriate pedestal.

these were prepared during my three years' partnership with Mr. F. C. Withers, to whom I am indebted for much valuable assistance in the preparation of this work.

It is not unfrequently said that architects' designs cost, in execution, more money than their employers are led, in the first instance, to believe will be necessary; but these assertions are for the most part ill-grounded, and arise from there being, here as elsewhere, a class of employers who profess to want much less than they really require, and who positively assert that they need about half of what they are determined to have. Such persons easily find a corresponding class of designers, and, of course, are always disappointed, as they richly deserve to be; but reasonable men, who are prepared to bring to the subject of spending their money the same good sense that has enabled them to realize it, find no difficulty in arranging their outlay in accordance with their wishes. For example, some of the houses in this volume have been very handsomely finished, and have cost not only much more than the outlay originally proposed, but much more than was necessary to complete them in a simple, rural manner. In no case, however, was the additional expenditure a source of dissatisfaction to the parties interested; such designs were carried out under the immediate inspection of their owners, and the desire for finish and refinement in detail increasing as the work proceeded, these gentlemen were well satisfied to enlarge, by degrees, their original intention as to cost. Some of the plans, on the other hand, have been executed for the exact sums specified in the contract; in these cases the proprietor, having approved of the drawings and specifications, has entirely ceased to interfere in the matter, except to pay the contractor's instalments when they have become due from time to time; and it may be stated, without any hesitation, that there are no insurmountable barriers to exactitude of estimate except loose instructions from the employer to his architect, and indefinite arrangements between the employer and his mechanics; both of which a proper amount of care at starting can readily prevent.

THE DOWNING MEMORIAL
ERECTED AT
WASHINGTON.

PREFACE
TO THE SECOND EDITION.

A new edition of this volume being called for at the present time, the opportunity has been taken to make some slight changes and omissions, and to introduce a few additional illustrations, with the descriptive text necessary to make them intelligible; but it has not been thought advisable to attempt any thing beyond this, as the work is essentially of a fragmentary and transitional character, and it would be impossible, with any amount of labor in revision that the author could now bestow, to give it a more satisfactory or permanent form.

A thoroughly comprehensive text-book on the subject of Rural Art, conceived and executed in the true spirit, would be a valuable addition to American literature; but some time must, in all probability, elapse before such a work can be successfully attempted, and, in the interim, its place must be, to a slight extent, supplied by works like the present, that merely seek to embody, in a shape fit for general reference, the thoughts and experiences for the time being of individual architects.

The six years that have passed away since this work was first published have not been of a character to suggest any rapid advance in popular taste with regard to the fine arts, and yet, even during this short period, marked at its commencement by wide-spread financial embarrassment, and closing in the midst of a terrible and exhausting civil war, some advance in the right direction seems to have been made. There is an increase in the demand for works of art of a superior class, and an important social idea has been developed in the "Artists' Receptions" that have now become so popular. Large public pleasure-grounds, such as the Central Park and the Baltimore Park, have been successfully established, an American Institute of Architects has been formally organized, and art papers of some special interest, such as the "Greek Lines," have found their way into the current popular literature.

There is, on the other hand, so far as Architecture is concerned, much cause for discouragement, and immense sums continue to be lavished with careless indifference on ugly, ill-planned buildings in every part of the country. In the course of the next decade, however, some more decided progress may be looked for; and among the refreshing signs of the present time is the fact that an association of young men has, during the past winter, been formed for the advancement of "*Truth in Art.*" The members hold that "all great art results from an earnest love of the beauty and perfectness of God's creation, and is the attempt to tell the truth about it." They also believe that, in all times of great art, there has been a close connection between Architecture, Sculpture, and Painting; that Sculpture and Painting, having been first called into being for the decoration of buildings, have found their highest perfection when habitually associated with Architecture; that Architecture derives its greatest glory from such association; therefore that this union of the arts is necessary for the full development of each. And this brief extract from their articles of organization indicates the spirit in which these artists expect to work, and is sufficient to show that much good may result from their earnest efforts in behalf of the good cause they advocate.

> "We stride the river daily at its spring,
>   Nor in our childish thoughtlessness, foresee
> What myriad vassal streams shall tribute bring,
>   How like an equal it shall greet the sea.
>
> "Oh, small beginnings, ye are great and strong,
>   Based on a faithful heart and wearyless brain!
> Ye build the future fair, ye conquer wrong,
>   Ye earn the crown, and wear it not in vain."

NEW YORK, *March* 23d, 1863.

B

STUDY FOR IRON TERMINAL, CENTRAL PARK, N.Y.

CONTENTS.

PREFACE.
                                                                       PAGE

Remarks on the Employment of Architects... ix

VIGNETTE.
Design for the Downing Memorial at Washington xiv

PRELIMINARY CHAPTER.
On the Design, Construction, and Detail of Country Houses.................... 25

VIGNETTE.
Design for a Village School-house.. 25

VIGNETTE.
Design for a Country Church.. 118

DESIGN No. 1.—(V. & W.)
A simple Suburban Cottage......................STUDY............................. 120

VIGNETTE.
Design for a Log-house..........................STUDY............................. 128

DESIGN No. 2.
A small rural Double Cottage....................STUDY............................. 130

VIGNETTE.
Design for a Hooded Door.......................STUDY............................. 132

DESIGN No. 3.
A Suburban Cottage..............................STUDY............................. 134

VIGNETTE.
Design for a partially inclosed Veranda..........NEWBURGH, N. Y.......... 138

DESIGN No. 4.
A Rural Cottage..................................FISHKILL, N. Y............... 140

VIGNETTE.
Design for a Rustic Outbuilding..................NEWBURGH, N. Y.......... 144

CONTENTS.

DESIGN No. 5.—(V. & W.)

A Suburban House .... NEWBURGH, N. Y. 146

VIGNETTE.
Design for a Double Suburban House... .... STUDY 150

DESIGN No. 6.

A Model Cottage .... STUDY 152

VIGNETTE.
Design for a Farm-house .... STUDY 158

DESIGN No. 7.

A Cottage Residence .... GOSHEN, N. Y. 160

VIGNETTE.
Design for an Artist's Studio .... RONDOUT, N. Y. 168

DESIGN No. 8.

A small Country House with Kitchen Wing ...... NEWBURGH, N. Y. 170

VIGNETTE.
Design for a Garden Outbuilding .... STUDY 174

DESIGN No. 9.

An Irregular Brick Country House .... YONKERS, N. Y. 176

VIGNETTE.—(V. & W.)
Boat Landing in the Central Park .... NEW YORK CITY 178

DESIGN No. 10.—(V. & W.)

A Suburban House with Attics .... NEWBURGH, N. Y. 180

VIGNETTE.—(V. & W.)
Design for a Garden Fence .... NEWBURGH, N. Y. 186

DESIGN No. 11.

A nearly square Suburban House .... RONDOUT, N. Y. 187

VIGNETTE.
Design for a Square House .... NEW HAVEN, Conn. 188

DESIGN No. 12.

An Irregular House without Wing .... SPRINGFIELD, Mass. 189

VIGNETTE.
Design for a small Country House .... ORANGE, N. J. 190

CONTENTS.   xxi

DESIGN No. 13.
                                                    PAGE
A Wooden Villa with Tower and without Attics...RAVENSWOOD, N. Y.......... 192

VIGNETTE.
Design for a Fence and Gate.......................RAVENSWOOD, N. Y.......... 196

DESIGN No. 14.—(D. & V.)
A Symmetrical Country House....................NEWBURGH, N. Y........... 198

VIGNETTE.—(D. & V.)
Design for a Coach-house and Stable ....NEWBURGH, N. Y.......... 202

DESIGN No. 15.—(V. & W.)
A Brick Villa with Tower and without Attics....STUDY....................... 204

VIGNETTE.
Design for an Observatory ....MIDDLETOWN, N. Y......... 208

DESIGN No. 16.
A Picturesque Symmetrical House ....NEWBURGH, N. Y........... 210

VIGNETTE.
Design for a Suburban Garden.....................NEWBURGH, N. Y.......... 214

DESIGN No. 17.
An Alteration of an Old House..................... NEWBURGH, N. Y........... 216

VIGNETTE.
Design for altering a Common Cottage Roof......NEWBURGH, N. Y........... 222

DESIGN No. 18.—(D. & V.)
A Picturesque Square House....................... NEWBURGH, N. Y 224

VIGNETTE.
Design for an Ornamental Garden ....NEWBURGH, N Y........... 230

DESIGN No. 19.—(D. & V.)
A Suburban Villa....................................GEORGETOWN, D. C........ 232

VIGNETTE.
Design for an Oak Mantle-piece....................FISHKILL, N. Y............ 236

DESIGN No. 20.—(D. & V.)
A Villa Residence with Curved Roof.............STUDY........................ 238

VIGNETTE.
Design for a Stable and Coach-house ....STUDY........................ 244

CONTENTS.

DESIGN No. 21.—(V. & W.)
An Irregular Wooden Country House .... Worcester, Mass. 246

VIGNETTE.
Design for a Gable Termination .... Newburgh, N. Y. 250

DESIGN No. 22.—(D. & V.)
A Suburban House with Curved Roof .... Newburgh, N. Y. 252

VIGNETTE.—(D. & V.)
Design for a Dormer-window .... Newburgh, N. Y. 256

DESIGN No. 23.
A simple Picturesque Country House .... Moodna, N. Y. 258

VIGNETTE.—(V. & W.)
Design for a small Rural Cottage .... New Windsor, N. Y. 268

DESIGN No. 24.—(V. & W.)
An Irregular Brick Villa .... Study 270

VIGNETTE.
Design for a Chimney with Ventilating Flues ..... Staatsburg, N. Y. 274

DESIGN No. 25.—(V. & W.)
A Suburban House with Curvilinear Roof .... Study 276

VIGNETTE.—(V. & W.)
Design for a Square House .... Study 280

DESIGN No. 26.
A Wooden Villa with Tower and Attics .... Study 282

VIGNETTE.
Design for an Entrance-gate and Piers .... Newburgh, N. Y. 286

DESIGN No. 27.
Family Cottage in the Mountains .... Study .... *Frontispiece.*

VIGNETTE.
Rustic Bridge .... Central Park, N. Y. 288

DESIGN No. 28.—(F. C. W.)
Stone Country House with Brick Dressings .... Clinton Point, N. Y. 291

VIGNETTE.
Shaded Seats .... Central Park, N. Y. 292

CONTENTS.

DESIGN No. 29.
Wooden Villa with Curved Roof .... GREENWICH, Conn. 294

VIGNETTE.
Square Villa with Curved Roof .... STAATSBURG, N. Y. 296

DESIGN No. 30.—(D. & V.)
A Villa of Brick and Stone .... STUDY 298

VIGNETTE.
Designs for Window-hoods .... STUDY 302

DESIGN No. 31.
Picturesque Stone Country House .... STAATSBURG, N. Y. 304

VIGNETTE.
Design for a Farm Cottage .... STAATSBURG, N. Y. 310

DESIGN No. 32.
An Irregular Villa without Wing .... STUDY 312

VIGNETTE.
Design for a Boat-house .... STUDY 316

DESIGN No. 33.—(V. & W.)
A Picturesque Villa with Wing and Attics .... STUDY 318

VIGNETTE.—(V. & W.)
Design for the Kitchen Wing .... STUDY 320

DESIGN No. 34.—(V. & W.)
A Town House .... NEW YORK CITY 322

VIGNETTE.—(V. & W.)
Design for a roomy Country House .... STUDY 326

DESIGN No. 35.—(D. & V.)
A Marine Villa .... NEWPORT, R. I. 328

VIGNETTE.—(D. & V.)
Design for a Coach-house and Stable .... NEWPORT, R. I. 330

DESIGN No. 36.
Irregular Stone Villa .... HIGHBRIDGE, N. Y. 332

VIGNETTE.
Coach-house and Stable .... HIGHBRIDGE, N. Y. 334

xxiv CONTENTS.

DESIGN No. 37.—(D. & V.)
		PAGE
An Irregular Stone Villa with Tower	STUDY	336

VIGNETTE.—(V. & W.)
Design for a small Country House .... STUDY 340

DESIGN No. 38.
Marine Villa with Tower .... NEWPORT, R. I. 342

VIGNETTE.
Coach-house and Stable............................. NEWPORT, R. I. 344

DESIGN No. 39.—(D. & V.)
A Villa on a large Scale............................ STUDY 346

VIGNETTE.—(V. & W.)
Design for a Grave-stone .... NEWBURGH, N. Y. 348

PRESBYTERIAN CHURCH, NEWBURGH. F. C. WITHERS, ARCHITECT.

VILLAGE SCHOOL.

VILLAS AND COTTAGES.

ALTHOUGH there is a cheering prospect for American architecture in the good time coming, its present appearance is in many ways far from satisfactory. Over the length and breadth of this country are scattered cities and villages by thousands, and public and private edifices innumerable; and yet we may fairly say, There are the buildings, but where is the architecture? There is the matter, but where is the manner? There is the opportunity, but where is the agreeable result? Is it in the churches? A few really creditable specimens may be pointed out, but the large majority are unquestionably deficient in truthful dignity and artistic beauty. Is it in the public buildings? Several fine works of art may at once occur to the mind, and although a floating doubt somewhat questions the Americanism of their expression, still, as they are nobly conceived and do not shrink from the ordeal of the artist's pencil, it is granted that they are successful. Then comes the question of the great majority again.

Does the memory linger with pleasure over the reminiscences of a provincial tour, and delight to recall the pleasant impression left on the mind by each elm-shaded town, with its tasteful hall, school-houses, library, theatre, museum, banks, baths, courts of justice, and other buildings cheerfully erected and gracefully arranged by its free and enlightened inhabitants —for their own use and pleasure, of course, but with a wise regard for mutual advantage and individual enjoyment, that insures the sympathy of every passing stranger; the more readily, too, as each discovers that he, even he, has been thought of, and that some study has been expended to give him pleasure? No, this is not the result to be looked for at present. Does the secret of beauty lie in the private buildings, the stores, the warehouses, the mansions, the villas, the hotels, the streets, or the cottages? There are probably as magnificent hotels and stores in the large cities of America as any where in the world. Architecture, within the last ten years, has managed to get a genuine foothold in this department of building; it has begun to *pay*, and that is an excellent sign, and one that offers food for reflection and solid encouragement; yet it is the few and not the many, even here, that speak of refinement, and a love of grace, which is as averse to meretricious display as it is to ungainly awkwardness. Among the private residences a great number are excellent; but still the mass are unsatisfactory in form, proportion, color, and light and shade. What is the reason of all this? Why is there comparatively so little beauty in American buildings? Some will say America is a dollar-loving country, without taste for the arts; others, that expense is the obstacle, and that the republican simplicity of America can not afford the luxury of good architecture. The latter of these solutions is clearly incorrect, for it is knowledge, and not money, that is the chief source of every pleasurable emotion that may be caused by a building. Indeed a simple, well-planned structure costs less to execute, for the accommodation obtained, than an ill-planned one; and the fact of its being agreeable and effective, or otherwise, does not depend on any ornament that may be superadded to the useful and necessary forms of which it is composed, but on the arrangement of those forms themselves, so that they may balance each other and suggest the pleasant ideas of harmonious proportion, fitness, and agreeable variety to the eye, and through the eye to the mind. All this is simply a matter of *study before* building, not of additional *cost in* building. The other solution of the problem, that Americans do not appreciate the beautiful, and do not care for it or value it, is a more specious but equally erroneous one. There are, doubtless, many obstructions that have hindered, and do hinder, the development of correct taste in the United States. The spring, however, is by no means dry, although these obstacles prevent its waters from flowing freely; and there is, in fact, no real difficulty that earnestness and ordinary patience may not overcome. One important evidence of a genuine longing for the beautiful may be at once pointed out. Almost every American has an equally unaffected, though not, of course, an equally appreciative, love for "the country." This love appears intuitive, and the possibility of ease and a country place or suburban cottage, large or small, is a vision that gives a zest to the labors of industrious thousands. This one simple fact is of marked importance; it shows that there is an innate homage to the natural in contradistinction to the artificial—a preference for the works of God to the works of man; and no matter what passing influences may prevent the perfect working of this tendency, there it exists; and with all its town-bred incongruities and frequently absurd shortcomings, it furnishes a valuable proof of inherent good, true, and healthy taste. Moreover, the greater includes the less. An actual love for nature, however crude it may be, speaks clearly of a possible love for art.

Till within a comparatively recent period the fine arts in America have been considered by the great bulk of the population as pomps and vanities so closely connected with superstition, popery, or aristocracy, that they must be eschewed accordingly, and the result is not *altogether* undesirable, though it has appeared to retard the advance of refinement and civilization. The awakening spirit of republicanism refused to acknowledge the value of art as it then existed, a tender hot-house plant ministering to the delights of a select few. The democratic element rebelled against this idea *in toto*, and tacitly, but none the less practically, demanded of art to thrive in the open air, in all weathers, for the benefit of all, if it was worth any thing, and if not, to perish as a troublesome and useless encumbrance. This was a severe course to take, and the effects are every where felt. But, after all, it had truth on its side; and candor must allow that no local, partial, class-recognizing advance in art, however individually valuable its examples might have been, could, in reality, have compensated for the disadvantage that would have attended it. Now, every step in advance, slow though it be, is a real step taken by the whole country. When we look at the ruins of old Rome, we say, What a great people! what temples! what mighty works! and undoubtedly Rome was really great *in individuals;* very great in a strong and clever minority, who spent with marked ability the labor of the weak and ignorant majority; but the *plebs,* the unlettered, unthought-of common people, the million, were not great, nor were they taught to be so, and therefore Rome fell.

During the last hundred years there has been a continuous effort to give to the American million the rudiments of self-reliant greatness, to abolish class legislation, and to sink the importance of individuals. "*Aut America aut nullus*"—"America or no *one,*" has been, is, and will probably ever be the practical motto. It is not surprising, then, that the advancement in the arts has been somewhat less rapid than the progress in commercial prosperity and political importance. The conditions were new, and, it must be confessed, rather hard. Continuous ease and leisure readily welcome art, while constant action and industry require time to become acquainted with its merits. To the former, it may be a parasite and yet be supported; to the latter, it must be a friend or nothing. The great bulk of money that is laid out on building in the United States belongs to the active workers, and is spent by them and for them. The industrious classes, therefore, decide the national standard of architectural taste.

The question then occurs, How is this universal taste to be improved? There is the sound, healthy material, unprejudiced, open to conviction, with a real though not thoroughly understood desire for what is good and true—there is plenty of prosperity and opportunity, plenty of money and industry, plenty of every thing but education and the diffusion of knowledge. This language may seem inapplicable to America, to whom humanity is indebted for the successful introduction of the common school system, which lies at the root of every healthy idea of reform now at work in the world, but is, nevertheless, true. The genius of American art may, with justice, say of the genius of American education:

> "If she be not so to me,
> What care I how fair she be!"

Education must be liberal and comprehensive as well as universal and cheap, or the result will remain incomplete. To secure any thing permanently satisfactory in the matter of architecture, professors of ability, workmen of ability, and an appreciative, able public are necessary. It would seem that the architects practicing in America are not at present, in the majority of cases, born or bred in the United States. They have, therefore, to learn and unlearn much before the spirit instilled into their designs can be truly and genuinely American. There is no good reason now why this state of affairs should continue. Architecture is a profession likely to be in considerable request here for several hundred years at least, and the demand is steadily increasing. Why, then, should not parents speculate for their sons in this line? Why should not the article, as it is for home consumption, be raised at home? It is an honorable calling; not certainly offering such splendid fortunes as the merchant *may* realize, but it is a fair opening, and the only capital that it requires, beyond brains and industry, is the expense for books and an education. When a fair share of Young America enters upon this study heart and soul, as a means of earning an independent position, we may expect a rapid, natural development of the architectural resources of the country, and that the present meagre facilities for artistic education will be gradually increased; the schools and colleges, also, will probably be induced, after a time, to include in their course of study subjects calculated to discover and foster, in the rising generation, such natural gifts as have a bearing on these and similar matters.

To insure workmen of ability, a reasonable chance to improve is the chief thing wanted. So long as the general demand is for monotonous, commonplace, stereotyped work, the average of ability will necessarily be low; but with opportunity, good, cheap, illustrated works, and a spirited weekly paper devoted to the special discussion of the subjects interesting to architects, engineers, carpenters, masons, and all the other trades connected with building—a paper that would diffuse sound theoretical and practical information on the art in general and in detail throughout the whole country, the advance would be rapidly felt; for wherever there is an American, there at least, be he rich or poor, is a reader, a thinker, and an actor. Self-supporting schools of design for painters, decorators, modelers, carvers, paper-stainers, etc., must follow in due course, for the positiveness of the need would soon become evident, and the object would then be almost gained. With reference to the appreciative and able public, the press is the improving power that is to be mainly looked to. Cheap popular works on architecture in all its bearings, popular lectures, popular engravings—and hundreds of them, and yet all good—these are the simple, truthful, and effective means that are to influence the public, by supplying a medium through which it may see clearly, and thus be led to criticise freely, prefer wisely, and act judiciously. Every year offers proofs of an advancing interest in this subject, and shows an increasing desire to respond to it in newspapers, magazines, books, etc., while the public is certainly not slow to buy and read.

The truth is, not that America is a dollar-worshiping country, with a natural incapacity to enjoy the arts, but a dollar-making country, with restricted opportunities for popular, artistic education, *as yet;* but when this want is freely ministered to, in the spirit that it may be, and it is hoped will be, ere long, there is every reason to conjecture that correct architectural taste may be as generally diffused throughout the United States as we at present find the idea of a republican form of government. We shall *then* hope for genuine *originality* as well as intrinsic beauty in American buildings; and this interesting subject of originality is, perhaps, worthy of a separate analysis and consideration.

In the United States it would seem that diversities of style and strong contrasts of architectural design are a perfectly natural occurrence, when we take into account the early history of the nation and the circumstances under which it sprung into its present prominent position. Differences of manner should, therefore, be contemplated without any troublesome sense of inconsistency being awakened, for such a charge would hardly apply with justice to results so clearly inevitable. The art of building faithfully portrays the social history of the people to whose needs it ministers, but can not get beyond those boundaries. We must remember, therefore, that principles of action, perceptions, convictions, habits of thought, and customs are the directors of all architectural design, and that wherever and however it may exist, it is one of several national exponents, not an independent affair with a cut-and-dried theoretical existence. Good architecture of some kind must spring up in any society where there is a love of truth and nature, and a generally diffused spirit of politeness in the ordinary habits of thought. Wherever, on the other hand, there is a wide-spread carelessness as to the development of the refined and gentle perceptive faculties, there inevitably must be a monotonously deficient standard of existence, and very paltry architecture as a necessary consequence; for the senses being deadened by inaction or abuse, poor seeing, hearing, smelling, tasting, and feeling, naturally result, and are reflected in the art of building, which exists entirely by supplying the demands of the bodily organs, and always shows whether they are vulgar, uncontrolled masters, galled serfs, or gay, active workmen. It is, moreover, an art so constantly before us, in some form or other, that it can not help being a friend or enemy to the improvement of civilized beings all the days of their lives.

The individuality of the American people does not appear to depend extensively on derivation or tradition, but on the character of the institutions by which it is surrounded, and on the elasticity of action that ensues. It is a people composed of many differing elements, but these are all exposed to a fusing power so strong, and so incessantly at work, that a single generation is often sufficient to bring into marked prominence the latent sentiments and springs of action that constitute the individual part of the national character. The settler may, to the last, be somewhat divided in his opinions, but the settler's son is sure to be an American, as far as *politics* are concerned, although at this point the active influence on him of the new country may appear to cease, leaving *social assimilation* to come about much more slowly.

Each of the European nations that have contributed to the population of this country, has, in its religious and domestic character, distinctive peculiarities and preferences, harmless in themselves, so far as others are concerned, and of comparatively private interest. These take their chance of life in the new country unmolested. The press having, of course, a gradual influence over them, while the national habit of traveling, by offering opportunities for tacit observation and change of opinion, without loss of self-complacency, is constantly at work rubbing down the rough edges of egotism, and rounding off the hard angles of prejudice. This influence every day enlarges its sphere of action, and will, doubtless, help a good deal to clear away the obstructions that at present hamper the *social* progress of the spirit of republicanism. Here lies the root of the matter; for whenever this spirit is permitted to flow freely into its natural channels, without being dammed up into an exclusive political mill-stream, it must lead to considerable social unity, and we may then, *but not till then*, look for the exercise of a power of fusion in manners and arts equal in its grasp to the one now almost omnipotent in politics.

The religious convictions of every country have, necessarily, a highly important influence over its social advancement; and America, so far as art is concerned, has received, till lately, nothing but blows from this quarter. Meagre sectarianism and private intolerance, under the names of religious freedom and universal toleration, have been serious drawbacks. However, the respect hitherto paid to mere formalism is now on the wane, and something more life-like is demanded—loudly at intervals, silently always. Besides the generally prejudicial effect that has thus been produced, there exists, here and there, a more distinct opposition to artistic grace and elegance on the ground that they are useless luxuries; but this sentiment is so impious, and the punishment daily inflicted on it is so sad, that it ought to be thoroughly exploded. Every sect agrees that there can be but one Creator, therefore all our created organs, sensations, and capacities must emanate from this fountain-head, and be intended for use; for if they have another source, there must be two first causes; and if they are not intended for use, the power exercised in their construction must be absurdly employed. Of course, either of these assertions would be highly irreverent. The fact is, that the evil influence at work in the world can mar many things, but make nothing; it can invert opportunity, misapply means, overdo or leave undone, and thus produce unpleasant, because unnatural results, but it can originate absolutely nothing: all its lying is spoiled truth, all its ugliness spoiled beauty; it can not help being in every instance secondary and negative: therefore, the moment we arrive at any thing so distinctly positive as a sensation or perception, we may know at once that it is irrational on any pretended ground of morality to "hide the talent," or oppose its healthy exercise. Earnest vitality is the social and artistic need in America. Every good thing, originality included, may be anticipated from that, for wherever it exists it bears fruit a hundred-fold; and the results of its influence on commerce show what may be expected when it pulsates as vigorously in the heart as in the pocket of this great republic. Cash will then assume its proper position, and money-spending will become the test of a man's ability instead of money-making. The arts generally, and architecture particularly, may then look with confidence for a better, because more natural era. It is not uncommon to hear it asserted, without much dissatisfaction, that in America one generation makes a fortune, the next squanders it, and the third or fourth, beginning afresh, at the lowest round of the ladder, amasses, by persevering industry, a new store of wealth, which is again run through as before. This melancholy fact is supposed by some to be agreeably connected with the abolition of the law of primogeniture, or the healthy action of political freedom and equality; whereas it is entirely contrary to the progressive spirit of true republicanism, and exists as a wholly unsatisfactory result of sheer neglect, ignorance, and waste. It has been reiterated over and over again, and tends to encourage the very incorrect and crude notion that a man, to be worth any thing, must begin with nothing, and make his own way by dogged perseverance at one idea for twelve or fourteen hours a day, through a series of years. Now, although such a man will probably be successful (for it would be hard if he were to miss his "pile," considering the sacrifice that has been made for it), yet money, being only condensed mechanical labor, can neither buy the capacity to enjoy the work it represents, nor the wisdom to spend one cent well. This is another affair altogether, depending on the exercise of a natural birth-right that may easily enough be parted with, but which it really seems poor policy to sell for a mere mess of dollars. Industry, energy, and perseverance make excellent servants, and deserve the respect due to good tools when sensibly exerted; but the unconditional acceptance of the stereotyped idea that man is a hack, and life a tread-mill, is certainly no mark of either wisdom or virtue.

There can not, indeed, be a more unpleasant spectacle than to see active, intelligent men, with long faces and knit brows, incessantly sacrificing time, health, home, and peace of mind to the one old "Moloch"—*business*, as if perpetual imprisonment were too good for reprobates, and business must, therefore, be converted into a portable bastille for the use of honest men. Every father, whatever may be his position in life, should undoubtedly use his best endeavors to enrich his children, but not chiefly with money. He should rather aim to start each one from a higher point of industrious, liberal civilization than he himself commenced at, and strive to relieve him from the difficulties that obstructed his own path. The exercise of such a spirit of foresight and progress would soon lead to artistic results worthy of the nineteenth century. It is worth remembering, too, that no occupation need be undignified, no labor graceless. Adam worked as a gardener, Franklin as a printer, Paul as a rope-maker, Æsop as a slave of all work, and Jesus Christ, at whose name it is said that every knee shall bow, as a carpenter. It is clear, therefore, that hard, manual labor is in no way removed from the highest developments of social philosophy and intellectual advancement. Every artisan, cultivator, or trader, may, if he think fit, not only be an honest, industrious republican, but a thoughtful, noble, and refined worker. All is within his easy reach. He has but to put forth his hand and pluck the fruit. The tree that bears it was planted by his ancestors, and is now daily tended.

though perhaps unconsciously, by himself; and this wide-spread appreciation of the possibilities that are within reach of every class, this all-encircling civilization, or an approximation to it, is absolutely necessary before art can take another healthy stride in advance. Galvanized action is worthless, however smartly it may be got up; there must be genuine life-blood flowing through all the members, freely and vigorously, or nothing good will be achieved; for the whole is made up of all its parts, and the parts of architecture, for instance, are, practically, the trades directly connected with it. The resources, therefore, of each of these must be well developed in detail before a really complete result can be arrived at. It is, moreover, well worth while to consider the large number of pursuits connected with this art. All the quarry-workers, lime-burners, brick-makers, lumber-merchants, glass-manufacturers, sawyers, masons, stone-cutters, plasterers, carpenters, joiners, tin-men, gas-fitters, plumbers, glaziers, iron-mongers, painters, carvers, gilders, modelers, decorators, architects, live by the constant demand for the exercise of this art of building; and whatever may be the pressure from without, it is only as these trades improve that the art improves.

In this country the wages of mechanics are good, and it is to be wished that they may ever remain so; but it is not equally desirable that the style of labor performed for those wages shall continue the same. On the other hand, it is to be hoped, and reasonably, when the subject becomes one of steady public interest, that the intelligence, skill, and taste of mechanics may be constantly heightened and improved, so that in time to come average ability may do easily and cheaply what is now considered superior out-of-the-way work.

All would be the gainers and none the losers by this advancement; for without the increase of wages each man would be able, with the same amount of personal effort as formerly, to purchase from his neighbor more positively valuable results. It is palpably evident, indeed, that a high standard of life can be quite easily attained by the working-classes in America, that is to say, by all whose circumstances render it necessary that they should do something for a living. Still there is another section of the community to be provided for—the born rich. Individuals in this predicament, in some parts of the world, have a gratifying position at once accorded them on account of their property, but this is far from being the case in America. There is a great deal of toil and consequent wealth in the United States; still, it is money-making, not money made, that commands respect. The science of spending is imperfectly understood, and the unsatisfactory results are apparent enough; but the idea of a moneyed aristocracy is every where repelled at heart with a scorn so contemptuous that it can scarcely be called indignant. A dilemma springs up from this state of things. Idleness is abhorred by successful men; they insist, therefore, on their sons becoming lawyers, or doctors, or going into business. Then follows a failure, in the majority of cases; for the spur to exertion that makes such pursuits satisfy men, is, in these instances, entirely wanting, as pecuniary circumstances do not, in the least, require the effort. Rich Americans fear lest their offspring may be looked on as useless members of society, and the instinct that leads them to do so is well enough as far as it goes, but the natural independent comment on it all is, Why spend so much time in making and saving money if it is to be rather an incumbrance than otherwise to the next generation. The real difficulty, and it is a serious one, is the limited range allowed by custom to intellectual energy. It is neither fair to the individuals, nor to the society of which they are responsible units, that the sons of rich men should be tied down to one or two money-making pursuits; they ought to be in every department of literature, science, and art, not as dilettanti connoisseurs, but as earnest laborers, striving boldly for a higher national excellence than has yet been achieved. This is *their* proper post. Poor men can scarcely afford to occupy it. It is a glorious position, the only proper one for them to assume; and so long as they neglect it, so long will wealth be misunderstood and misapplied. The rich should study to be practical theorists, so that the less rich may be theoretical practitioners. Every young republican of means in America should aim to be *aristo*cratic in its literal sense; that is, to be "*aristos*"—the very best. He has advantages which his comrades have not. He can afford to give ample, unembarrassed study to any subject that suits his powers, and to work out its resources quietly and steadily. He should be one step in advance of the rest of creation, a leader in the foremost rank of the foremost band. The value of a class of men thus occupied would be unquestioned, and it would not be so unnatural then for a parent to labor for money, so that his son might enjoy the rightful opportunity to live an easy life of elevated action and noble exertion.

The nature of the pursuits men follow should be examined into and tested. As we are not good, we need preachers; as we are not straightforward, we must have lawyers; as we are not natural, we want doctors: and we are much indebted to all these gentlemen; but as we *are* virtuous, we ask for something besides, that shall be less negative and more actual. We demand poets, mechanics, philosophers, men of business, artists, authors, sailors, inventors. Possessing the capacities of all these in embryo, we ask, not that we may blindly admire the individual, but that we may fairly appropriate the spirit of his work, and be, when not laboring at our own specialty, at small cost, and in a quiet, general way, what each professor is in an individual, troublesome, and particular way. This seems to be the scope and intention of life with such a basis as American freedom; and exactly to the extent that it is recognized and acted on is the advance in art and science, and in all that makes the best part of life.

It is only in a state of society in which things are valued intrinsically, and not for what they will fetch, that any art can begin to progress or hope for a chance to become vital or original; and so long as the enjoyment of a purchase depends much on the opinion of others, and but little on its own merits, artistic invention is likely to remain at a stand-still. Taste must be real, unborrowed, and individual, to accomplish any thing; and even a small allowance of earnest perceptive conviction is better than any amount of follow-my-leader opinion. If excellent architecture can give innocent pleasure, it is certainly worth having, and all Americans ought to have it with as little delay as possible; not for the sake of gratifying any petty spirit of rivalry, or indulging any national or local pride, but simply because it is *worth* having; on the same principle that every healthy man ought to enjoy dining daily, not on account of his being able to afford richer food than his neighbors, or because he happens to know a dozen people who live poorly (for if he can derive any appetite from such facts he is no true man), but solely because his good constitution requires regular sustenance.

In America perfect liberty, that absolute essential for healthy life, has been, in due course, talked for, fought for, legislated for, and, in these Free States, decidedly realized; and it seems, therefore, scarcely fair now to train all the best men to be lawyers and politicians, because the talent is more wanted somewhere else. The sensation of freedom is nothing more than the felt certainty of non-interference, and, however complete it may be, it can neither supply the will to do any thing, nor suggest any deed to be done; it is like light, only perceptible when reflected from an object; it offers a solid rock on which to build, but not one idea for the superstructure adapted to it. In America this rock commands a boundless prospect, and no fitting or enduring edifice can be erected on it that does not include the most liberal manners, the most generous aspirations, the most noble institutions, and the most pure and beautiful arts that unfettered humanity is capable of conceiving. There has not, indeed, been, from the commencement of the world till this moment, an opportunity for the advance of the fine arts so replete with the material of true success as now exists in America; this advance is a question of choice, not time; of purpose, not ability; of direction, not force; there is *capacity* enough spread over all the country, and being wasted daily: it is *conviction* and *will* that are needed.

When the talent and energy that are fostered by American institutions are distributed with tolerable fairness, we shall, among many other things, be justified in expecting to find in every architectural effort, not something so new that it is unintelligible, but some distinctive characteristics that show it to be a genuine American invention. These, however, can hardly be expected to depend much on the employment of really new forms. Webster and Clay were orators of originality, but their words were all old. Their stock in trade is common property in the form of a dictionary, and the boundary lines, over which neither ever ventured to pass, are fairly set forth in a good grammar. Any desire on their part to invent a brand-new language would have been, of course, absurd, and any wish to produce a brand-new style of building is, without doubt, an equally senseless chimera.

All previous experience in architecture is the inherited property of America, and should be taken every advantage of. Each beautiful thought, form, and mode that is not unsuited to the climate and the people, ought to be studied, sifted, and tested, its principles elucidated, and itself improved on; but the past should always be looked on as a servant, not as a master.

Individual sentiment and education should be encouraged to act freely in every instance, and by degrees that important fact, a genuine public taste, will be fairly unfolded. The authority of precedents will then be unneeded, for actual ideas, such as "fitness," "unity," "variety," will give the critical standard to the general taste. Every individual of sound mind will then help to improve the national architecture, for each will resolutely refuse to admire any structure that does not seem agreeable to him or her individually, and all will freely insist on a right to call *good* whatever coincides with their untrammeled, but not uncultivated natural perceptions. Emerson says forcibly on this point, "Why need we copy the Doric or the Gothic model? Beauty, convenience, grandeur of thought, and quaint expression are as near to us as to any, and if the American artist will study with hope and love the precise thing to be done by him, considering the climate, the soil, the length of day, the wants of the people, the habit and form of government, he will create a home in which all these will find themselves fitted, and taste and sentiment will be satisfied also."

In a country like this, where the printing-press accompanies each stride that is made into new localities, and where every step is marked by a building of some sort, it seems inconsistent that there should be but little popular literature on architectural matters; yet such is undoubtedly the fact, and although Americans are certainly diligent readers and energetic builders, their habit of reading has scarcely had so much influence for good on their habit of building as might naturally be expected, when we consider the practical character and universally recurring interest of the subject of domestic architecture.

It has not, certainly, till within the last few years, been an easy matter to place before the public the necessary illustrations in a convenient form, and as mere verbal descriptions of plans or designs are seldom thoroughly intelligible, this difficulty has probably retarded the diffusion of popular architectural information. Now, however, with the present rapid development and general application of the art of wood-engraving in the United States, this hindrance no longer exists, and a fair field is open for the free communication of ideas among American architects, and for the profitable interchange of hints and suggestions.

The study of what has been done by other nations, though useful as a help, will never, by itself, lead to much result in America, where the institutions, the needs of the climate, and the habits of the people, have a distinctive character that requires special consideration; and this remark applies particularly to rural architecture. Thus the Greek mode, though completely beautiful when contemplated from a proper point of view, has for its leading characteristic a passionless repose that is not heartily sympathized with either by the American atmosphere or the spirit of this locomotive age; and, consequently, no architectural effort imitated from the Greek can help being, to a great extent, a mere lifeless parody. The failure is generally very conspicuous, but even in the least unsatisfactory instances some absurd inconsistency is sure to assert itself. Common sense will insist on chimneys and verandas, and the pure classic outline in due course suffers grievous mutilation, being thus punished for its intrusion into a locality where it had no business to be attempted.

Styles like the Chinese or Moorish assist us but little, though each exhibits isolated features that deserve careful examination. The Moorish, for example, shows what magical effects may be produced by light, recessed arcades, and gives some good suggestions for verandas. The Chinese again, with its trellises and balconies, is interesting in detail; but neither of these phases of architectural taste is of comprehensive value. They are very deficient in compactness and completeness of plan, and in artistic design they depend too much for their effect on delicate and elaborate ornamentation; such decorations as paneling, carving, painting, and gilding, may be readily enough obtained where a clever, industrious, efficient pair of hands can be hired for a few spoonfuls of rice *per diem*, but not so easily in a country where every one is as good as his neighbor "and better," and where ordinary mechanics ask and get two or three dollars for a day's work. The irregular Italian, and the later modifications of the Gothic, are the most useful types to analyze; but the flat-terraced roofs of the first have to be avoided on account of the snow, and the latter has to be adapted to the use of verandas before it can be acceptable. Nor is this all: there is in this country a perpetual necessity for compactness of plan, however large the house may be, because, as it is invariably difficult to get efficient servants, it is desirable to save labor in every possible way. In this particular neither the Italian nor the Gothic examples help us materially; they delight too much in halls and passages, long corridors and wide vestibules, galleries, and staircases. This sort of rambling arrangement does not answer here—the difficulties of heating and service render a closer attention to concentration desirable—nevertheless, a sufficient privacy, and a freedom from any appearance of meanness, is the right of every house, however small its scale. The English country houses and cottages have undoubtedly claims to our best consideration; but it is from an examination, by means of illustrations, of what is going on at home, called forth by the actual needs of people, more than from a study of foreign examples, that the general taste for architectural comfort and beauty in country houses is likely to improve. Any genuine step in advance will be responded to at once by the sympathetic perceptive faculties of individuals who may notice it, and the result, so far as it bears on their needs, will remain daguerreotyped in the memory. Whatever, on the other hand, has no reference to local habits and experience, will be passed over without receiving much consideration. Every active-minded man is in a position to understand and criticise such examples, and though they may have little of the pretension or extent that specimens of villa architecture in other differently constituted countries would afford, they will have the practical advantage of offering definite starting-points for farther improvement at home.

This would hardly hold good if there was very little doing; but such is not the case. There has been latterly an immense number of buildings of this nature going up in all parts of the United States; numberless villages have sprung into existence, and much thought has been given to the subject. A very transient visit into any part of the country shows, however, that most of the villas and cottages are erected without regard to artistic propriety, and at considerable loss to their owners from the useless outlay incurred by adopting ill-considered plans, and the subject, as well as the majority of the houses, would be improved by a little more ventilation. Square boxes, small and large, are springing up in every direction, constructed without any attempt at proportion, or the slightest apparent desire to make them agreeable objects in the landscape. These tell their tale simply and unceremoniously: they are the natural result of the migratory, independent spirit pervading the industrious classes in America, and offer interesting evidences of the genuine prosperity of the country, for they show not only that the landlord and tenant system is disliked, but that almost every store-keeper and mechanic can contrive, even when quite young, to buy his own lot and live in his own house. On the other hand, however, they demonstrate that the capacity for enjoyment, and the appreciation of what is really desirable in life, that should naturally accompany this active and successful industry, are wanting. Each of these bare, bald, white cubes tells its monotonous story of a youth passed with little or no cultivation of the higher natural perceptions, and of a system of education in which the study of the beautiful in its most simple elements is neglected and apparently despised. The lack of taste perceptible all over the country in small buildings, is a decided bar to healthy, social enjoyments; it is a weakness that affects the whole bone and muscle of the body politic; and it is a needless inconsistency, for a full exercise of freedom of speech and action should naturally result in a full, free exercise of the innocent enjoyment that unfettered industry renders possible, and a refined propriety and simple, inexpensive grace ought habitually to be the distinctive marks of every habitation in which a free American dwells.

Unfortunately, however, this is not the case. Even the village school itself, in which the earliest and most active germs of progressive thought are commenced, is almost universally a naked, shabby structure, without a tree or a shrub near it, and is remarkable chiefly for an air of coarse neglect that pervades its whole aspect. The improvement of the village school-house is probably the most powerful and available lever that can be applied toward effecting a change for the better in the appearance of rural buildings generally: all see it, all are interested in it, and all are more or less influenced by its conduct and appearance. It is placed under the control of the leading men in each place, and it might easily be made the most cheerful and soul-satisfying building in the neighborhood, instead of, as at present, a God-forsaken, forlorn-looking affair, that is calculated to chill the heart and insult the eye of every thoughtful beholder. The cost would be utterly incommensurate with the advantage to be obtained. An extra hundred or hundred and fifty dollars at first starting would do much. The roof might then have a good projection, and be neatly finished. Some sort of simple porch might be added, the chimney might be slightly ornamented, and the rest would then depend on proportion, color, and surrounding the building from time to time with shrubs, creeping vines, and young trees. These, in after years, would offer a welcome shade, and give an air of domestic comfort and liberal vitality to the whole effect. A similar result, through precisely similar means, would probably, in course of time, be arrived at in the small cottages in its vicinity, and, as success would be cheap and invariable, the example would have a fair chance of spreading. Such a simple, unpretending style of building as is sketched on page 25, admits of endless variety of design, and is within the reach of every civilized community.

Our subject having thus led us to the consideration of school-houses, it may be worth while to add a few words as to the schools themselves.

True and intelligent republicanism clearly points to a state of society in which the private possession of great pecuniary wealth ought to be a comparatively unimportant matter, because it should yield to its possessor but little more real comfort, or even luxury, than can be readily acquired by every industrious man. Com-

D

plete protection from the weather in healthy, well-ventilated, comfortably-appointed, and tastefully-arranged apartments; good food, scientific cookery, and an ample supply of artificial light; appropriate clothing, pretty furniture and draperies, delightful books, engravings, and works of art, may all be obtained at little cost by a skillful combination of liberal economy and wise management; and it will be found that the richest man in the world can scarcely realize more than this, though he may, of course, carry out the idea on a very magnificent scale. But even here the man of small means *may* be almost on the same footing as the millionaire, for public baths, gymnasiums, theatres, music-halls, libraries, lecture-rooms, parks, gardens, picture-galleries, museums, schools, and every thing that is needed for the liberal education of an intelligent freeman's children, can easily be obtained by the genuine republican if he will only take the trouble to *want* them. All, and more than all, of these sources of gratification lie folded up in his industrious palm. He may either clench his fist and fight his way through the world without allowing the treasure he holds to see the light, or sell it to his brother for a mess of pottage on the old aristocratic plan; or he may keep his birth-right himself, and cultivate it, as nature intended he should, for his own and the public enjoyment. A correct general choice in the matter must be the work of years, and can only result from a refinement in popular education that will urge the unspoiled, pliable young minds of the rising generation to the study of the beautiful as well as to the acquirement of reading, writing, and arithmetic. The leading principles of good taste should go hand in hand with the multiplication table, and every common school class should have its artistic as well as its literary and oratorical book of selections from the best authors for every day public reading.

One especial disadvantage that rural art labors under in America is, that the plans of country towns and villages are so formal and unpicturesque. They generally consist of square blocks of houses, each facing the other with conventional regularity; and this arrangement is so discordant with the varied outlines characterizing American scenery, that Dame Nature refuses, at the outset, to have any thing to do with them, and they never seem afterward to get any better acquainted with her. Except, perhaps, in a very large city, there is no advantage gained by this intense monotony of arrangement, and it is much to be regretted that in the many new villages that are being erected the same dull, uninteresting method is still predominant.

The great charm in the forms of natural landscape lies in its well-balanced irregularity. This is also the secret of success in every picturesque village, and in every picturesque garden, country-house, or cottage. Human nature, when allowed a free, healthy scope, loves heartily this well-balanced irregularity, and longs for it in life, in character, and in almost every thing. It is the possession of this same quality, even when the balance is incompletely kept, that makes the stirring, unconventional, free-spirited man so much more interesting and agreeable than the cold, correct, and somewhat unsympathetic gentleman who never does any harm to any one, and whose equanimity is never disturbed. We want far less formality and restraint in the plans of our new villages. The roads should wind in graceful, easy curves, and be laid out in accordance with the formation of the ground and the natural features of interest. A single existing tree ought often to be the all-sufficient reason for slightly diverting the line of a road, so as to take advantage of its shade, instead of cutting it down and grubbing up its roots. In a case that recently occurred near a country town at some distance from New York, a road was run through a very beautiful estate, one agreeable feature of which was a pretty though small pond that, even in the dryest seasons, was always full of water, and would have formed an agreeable adjunct to a country-seat. A single straight pencil-line on the plan doubtless marked out the direction of the road; and as this line happened to go straight through the pond, straight through the pond was the road accordingly carried, the owner of the estate personally superintending the operation, and thus spoiling his sheet of water, diminishing the value of his lands, and increasing expense by the cost of filling in, without any advantage whatever: for a winding road so laid out as to skirt the pond, would have been far more attractive and agreeable than the harsh, straight line that is now scored like a railway track clear through the undulating surface of the property; and such barbarisms are of constant occurrence. Points of this nature deserve the utmost attention, instead of the reckless disregard they generally meet with. When once a road is laid out, its fate is settled, and no alteration is likely to be made: it is, therefore, the more desirable that its direction should be well studied in the first instance.

If we now attempt to take a general view of the subject of domestic architecture, we shall find, in the first instance, that every residence intended to be satisfactory must be comfortably *planned*, pleasantly *designed*, and soundly *constructed*.

In any design that is intended to be used by an intelligent human being, the general distribution and detailed arrangement of the accommodation to be furnished, or what is called *the plan*, is the first point that should occupy our attention; for the most simple idea admits of a good or a bad arrangement. Let us take, for example, a house that is to consist of but one room, as in plan A. The door opens immediately opposite the fire-place: a cold draught is, therefore, likely to be constantly traversing the whole length of the floor of the apartment from the door to the fire, and as the chimney is placed in the outer wall, a great deal of heat will be lost. Moreover, the bed, C, and the sink, S, are entirely exposed to view, and thus privacy and cleanliness are scarcely possible. Now a man may, with comfort and decency, make his kitchen his living room; but he will find it disagreeable if he has, in addition, to use it as a bedroom and sink-room. It is therefore desirable to improve on this plan. The diagram B shows a different, and, in every way, a more sensible and convenient arrangement of the same space. Here the door and chimney are so placed with reference to each other, that the minimum of draught occurs in the room when the door is opened. The chimney is built in the body of the house, so that as much heat as possible is saved. A few feet of board partition set at the back of the chimney supplies a somewhat private recess for a bed, and also an entrance lobby with room for the sink; and the principal room and both recesses may communicate by

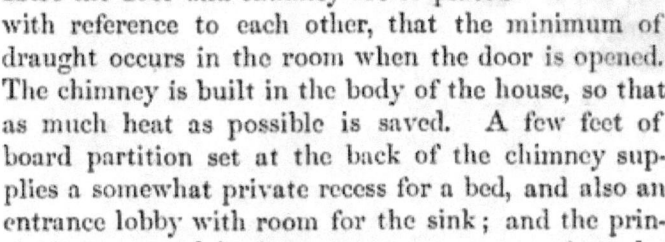

means of tin pipes through the ceiling with an air flue carried up alongside the smoke-flue, thus thoroughly ventilating the whole building. The two closets are placed at the other end of the room, so that a window seat, that may also be a locker to contain a supply of coal, can be arranged between them, thus rendering the interior appearance of the room more agreeable; and a strip of curtain, or, if thought worth while, a light door hung on each side of the chimney-breast will give privacy and an air of snugness to the whole arrangement. Now, this plan in execution would cost perhaps $10 more than the other; and taking the interest of this $10 at 10 per cent., the *cash* difference to the occupier would be $1 a year, while the *comfort* difference to any one with the slightest taste for that blessing would be incalculable—the one residence being inconvenient and vulgar, while the other, so far as it goes, is commodious and comparatively elegant.

This sketch of a plan is introduced merely for the purpose of illustrating the difference between an ill-considered and well-considered mode of working out on a very small scale the simplest possible idea of a decent human habitation; but the principles involved in its arrangement apply with equal force through the whole range of domestic architecture. The first thing that is needed is a kitchen; the second, an inclosed lobby. A separate bedroom is the next step for comfort; and we then advance to a plan that provides a living room separate from the kitchen, a hall with a staircase in it, and bedrooms up stairs. A house with this amount of accommodation should have a veranda attached to it. A separate staircase-hall and a second living room would be the next addition to comfort, and we may then proceed upward in the scale of accommodation to any extent that is required, adding separate rooms for special purposes, a servants' staircase, bath-rooms, etc. But before leaving this part of the subject, it seems desirable to say a few words as to the proper scope for the plans of country residences of moderate size.

By far too many of the villas that are built are extensive and costly, and many persons in easy circumstances are deterred from building a home in the country, because they are impressed with the idea that they must erect a large house or none at all. This prevalent feeling prevents a great deal of enjoyment of rural life that might otherwise be realized, and requires to be looked into and criticised.

All that appears to be necessary for real comfort in a villa or cottage residence, exclusive of the bedrooms and offices, is a parlor of tolerable size, which shall be the general living-room of the family, and another apartment contiguous to, or connecting with it, to be used as a breakfast and dining room. If a third large room, to be called either library or drawing-room, is required, the whole scale of the house is materially enlarged, and its cost much increased.

It has been, and is too much the custom, both in town and country houses, to consider the dining-room as a part of the house to be used solely for eating and drinking purposes, and to give it but little attention for that reason. It is, indeed, quite common to find, even in comparatively large houses, a meagrely-furnished apartment in the basement set apart as the scene of whatever daily festivity is carried on in the house.

If a country residence is built on sloping ground, so that the basement rooms on one front are entirely unobstructed, and are supplied with windows overlooking the garden, this objection is not so strong. But even then, the trouble of going up and down stairs to and from the sitting-room is annoying, and it is far preferable to have both rooms near together on the principal floor. But when, as is generally the case, the house is built on level ground, and the lower rooms are lighted solely by area windows, nothing can be more entirely opposed to the idea of freedom that is suggested by life in the country than a basement dining-room. It is in this apartment that the different members of the family are sure to assemble several times a day, though they may be almost completely separated at other times by circumstances, or the various pursuits that occupy their attention, and it is highly desirable that such a room should fitly and cheerfully express its purpose, and be one of the most agreeable in the house, so as to heighten the value of this constant and familiar reunion as much as possible, and to encourage in every way, by external influences, a spirit of refinement and liberal hospitality. The fact is, that the art of eating and drinking wisely and well is so important to our social happiness that it deserves to be developed under somewhat more favorable circumstances than is possible in a basement dining-room. There is no necessity in any country house that such a room should be restricted in its use to one purpose. If fitted up with book-cases, and enlivened by engravings, it will be constantly used as a family room, for, with proper pantry arrangements, it can be left entirely free in a few minutes after each meal.

Design No. 6 was prepared especially to illustrate these remarks.

When the plan proper is perfected, and all the accommodation determined on is arranged in the most economical and agreeable manner, there are still two main points that need attention, if the house, however small, is to be healthy and sweet. These are ventilation and drainage. Many volumes have been written on the subject of ventilation, and much rhetoric has been expended on the effort to show that lungs require fresh air. But although this is a self-evident proposition to any man, woman, or child who will give the subject a moment's consideration, it is lamentable to perceive that this moment's consideration is seldom given, and that the common practice of ordinary house-building is in opposition to plain sense so far as ventilation is concerned. Yet the leading points may be briefly stated. Any apartment, to be a healthy one, needs a current of air always passing through it, whether the windows are open or shut; and to insure this, every one must see that it is necessary to have an inlet and an outlet. Now an inlet is commonly provided by the warm air register from the furnace, which should be left open as a cold air register in summer, and is always insured more or less by the opening and shutting of doors. But the outlet requires special attention. It should be an orifice suited to the size of the room, opening into a flue that is not a smoke-flue, near the ceiling, for summer ventilation, or for use in colder weather when the apartment is crowded; and an orifice opening into a similar flue near the floor for ordinary winter ventilation: this latter great desideratum is provided to some extent in every room that is furnished with an open fireplace, whether a fire is ever lighted in the grate or not, and for this, if for no other reason, it is desirable that the open fireplace should never be omitted in rooms that are intended to be occupied either as living or sleeping apartments.

The objection that is generally raised against ventilation is, that it is uneconomical in winter, more fuel being needed to keep a room warm if there is a free escape for the air, than if its circulation is impeded; and this is granted at once, as far as the expenditure of *fuel* is concerned, but not as regards *economy* in any proper sense of the word. It would, of course, be a saving in money if a family should acquire the habit of eating spoiled meat, as such stuff could probably be bought cheap; but it would be very poor economy; and it is equally senseless for human beings to feed their lungs with spoiled air, on the plea of saving money by it.

Efficient drainage is an equally important part of any plan for a residence in the country, and some means of getting rid of the sewerage and waste water must be provided. A complete connected system of pipes communicating with a common sewer is not to be had in the country, and we must, therefore, resort to the plan of a cess-pool or manure tank. This should be placed at some distance from the house, and as far as possible from the well. If the premises are small, and it is necessary that the well should be nearer than 100 feet to the cess-pool, the latter must be cemented and rendered thoroughly water-tight, and emptied from time to time, or evil consequences may ensue. This receptacle should be connected with the house by brick drains, or, what is better, by four-inch or five-inch earthen pipes, if they can be readily procured. All these communications require to be properly trapped, so that there may be no continuous air-passage from the cess-pool to the house, or noxious gases will certainly rise through the pipes, and be extremely offensive in every way.

A water-closet, or its equivalent, is an absolute necessity in any house that is proposed to be a convenient and agreeable residence; but as there is sometimes a difficulty, and always an expense, in arranging a regular water-closet, it is desirable to invent some simple plan which shall approximate to its advantages at little cost. A necessary abutting on the house, or communicating with it by a covered way, if properly arranged, will answer the purpose. The diagram shows a plan that I have prepared and adopted with advantage in several cases. The idea is to form a comparatively small receiver, with a circular cone in the centre, immediately under the floor of the closet, both cone and receiver being of brick coated with asphalt, or some cement that will not be affected by frost. On one side this receiver is connected with a rain-pipe from the 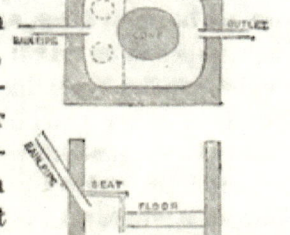 building, and through a proper trap on the other side, with a drain leading to the regular cess-pool. By this arrangement we have but a small surface exposed to the action of the atmosphere, and this is changed whenever there is any rain. The object of the cone is to reduce the surface and increase the effect of the wash, while the rain-pipe acts as a ventilator, carrying any gas that may be generated up to, and above, the roof of the main house. If the outlet on the roof is near a dormer-window, and it is consequently objectionable to ventilate at this point, the pipe may be constructed as

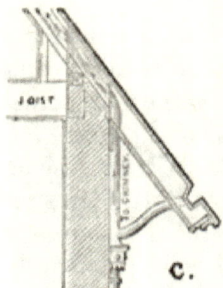

shown at C, and the ventilating tube carried up under the rafters into a chimney, where it can do no harm, while the increased height will insure a draught in the right direction.

When we consider the troublesome, unhealthy, indelicate, ugly effect of these outbuildings, as usually constructed, it is certainly worth while to consider this simple plan, which is applicable to any house in any situation, and can hardly get out of order practically, for when the exposed surface freezes up in winter, no harm is done, as there is nothing to burst, and no evaporation takes place in very cold weather; while, on the other hand, if a long drought occurs in summer, a few pails of water poured in from time to time will set matters right till a shower comes. Every water-closet, either external or internal, should have a fair-sized window in it, so hung that the upper half, at least, will move up and down.

It is a peculiarity of warm, soapy wash-water, that it very readily deposits its greasy particles in passing along a cold drain-pipe; and to avoid the inconvenience that otherwise ensues at short intervals of time, both from foulness and stoppage, it is desirable to construct a small, tight cess-pool near that part of the house in which the wash-trays or sink are situated—say three feet in diameter and twelve feet from the main wall, with a trap and overflow into the main drain, arranged as shown at D, on the opposite page, and with a piece of straight pipe into it that can be cleaned its whole length from the inside of the house by taking out an air-tight stopper and inserting a long stick and old cloth. It must also have a cover-stone at the top, and it is well to form a connection with a rain-water or other pipe for ventilation. By this arrangement the greasy water passes into the straight pipe, and thence into this small cess-pool, in which the greasy matter rises at once to the surface of the water, where it cools and coagulates harmlessly, while the mere dirty liquid passes under the trap and into the main drain. Once a year, perhaps, the pipe and small cess-pool should be cleaned out;

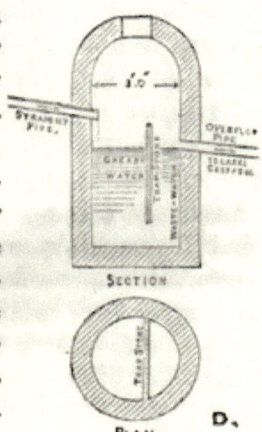

but it will be readily seen that this is a simple and short operation. From the omission to take some precaution of this kind, families are frequently annoyed both by bad smells from their drains (for this greasy, soapy deposit is more foul than any other) and by the inconvenience and expense of taking up and relaying long lengths of stopped-up drain-pipe every now and then.

Having thus briefly remarked on the plan or convenient arrangement of the accommodation, we proceed to the *artistic design* of rural buildings, particularly of their exteriors, and we must take care, at the outset, not to be deceived as to the true principles and laws that regulate this important part of the subject.

Architecture is entirely the invention of man, and, as it expresses his needs and his nature, it must necessarily be regulated by the laws to which he is subject. At the same time, it is equally clear that it can have no independent laws of its own, simply because it has no independent existence. As it seeks to please the eye, its forms and colors should be carefully designed in accordance with the laws of the eye, or it will be a failure, so far as this organ is concerned. As it addresses itself to the intellect, it ought to be orderly and without any appearance of accident in its conception, or it will appear unintellectual. As it appeals to the heart, it requires to be forcibly and artistically true in its expression, or it will remain a lifeless collection of mere building materials; and as it ministers to the soul, it must be beautiful and pure in its intention, or it will be ugly and baneful in its influence. It is always the mirror of its age, accurately reflecting the customs, morals, and science that prevail in any nation at a given period; and as these have been dissimilar at different times and places, architecture has naturally crystallized in various parts of the world into what we call separate styles. Still, we must never forget that the elaborate divisions that have thus sprung up, expressed by such words as Grecian, Roman, Gothic, or Hindoo, belong to the *history*, not the *art*, of architecture. The self-same geometry shows itself transparently in all styles, fashions, and orders. The prismatic colors are permanent facts. Human nature is to-day what it always was, and always will be, till man ceases to exist as man. There is, therefore, open to us, if we choose to adopt it, one broad, natural, open-air standard of criticism belonging to all architectural works, irrespective of style or fashion. And as this standard is simple and intelligible, it is to be preferred to any narrow, sectional rules dependent on the laws of this style, or the regulations of that order, or the requirements of some special professor. We may each if we choose to take the trouble, go to the fountainhead and decide for ourselves.

The points of climate and atmosphere require, in all countries, careful local analysis before the interior arrangement of any habitation can be successfully adapted to its purpose as a healthy, convenient residence. And they certainly demand no less study, though in another way, if its external appearance is to be judicious and tasteful. In the plan, indeed, each sense, in turn, has to be duly considered, while, in its artistic effect, but one is appealed to. Yet this one is the most important of all; for the light of the body is the eye, and it is to the eye, with the infinite host of progressive ideas, to which it acts as the mysterious portal, that the design of every building has the opportunity of artistically ministering.

Throughout the whole of nature we perceive a strong love for balance, every appearance of repose depending entirely on an equilibrium of antagonistic forces; and as this state of sensitive balance is the only natural condition of true life and joy in any exercise of the human faculties, the eye partakes of the universal desire, earnestly seeking for it in all examples of form and color, including light and shade, in all their varieties.

We may with propriety, so far as our subject is concerned, call this balance *proportion* when speaking of *form*, and *harmony* when speaking of *color*. Still, it is not sufficient that the various parts of a building should be in proportion to each other, or that it should be, as a whole, harmonious in its actual coloring. It must also possess these qualities when considered with reference to climate, scenery, and surrounding objects. One peculiarity of the American climate is an absence of humidity in the atmosphere. The weather is generally clear, and the pure, dry air is so transparent, that it permits a distinctness of outline to objects even at a considerable distance from the eye. This habitual freedom from moisture is not confined to any season of the year. We have, undoubtedly, misty, and even foggy days, and these occur not unfrequently in the transition from winter to spring; but for the greater part of the summer, and during the fall and cold months, the bright sun shines out week after week with little intermission. In Italy or the East the air also allows remote objects to be very clearly seen; but it is, at the same time, so suffused with an attenuated, almost imperceptible, hazy medium, that the direct, glaring rays of the sun are subdued and softened before they meet the eye, and a delicate gradation of perspective distance, with an agreeable variety of harmonious half-tints, is the natural result. In America this seldom occurs. The supply of light is usually free from any mellowing veil: it is, therefore, colorless or white, and very decided in its pictorial character. A few Indian Summer days in November give us, indeed, some exquisitely-beautiful opportunities for the study of vaporous, dreamy effects; but these are soon enjoyed, and pass away, leaving the ordinary, translucent, unclouded character of the atmosphere more apparent than ever.

The light in America being, therefore, powerful and somewhat trying to the eye, it seems desirable to select arrangements of form and color in rural architecture that will rather relieve than increase this fatiguing effect. It is a well-known fact that, if a person looks steadfastly for a length of time at any decidedly red surface, the next object that occupies his attention will have in it a tinge of green, no matter what its *real* color may be. Nature thus seeks to restore the equilibrium, and the strained organ is somewhat refreshed; but it will naturally be much better satisfied if the object happens to be of a cool, green tint, for the balance will then at once be rapidly and agreeably re-established. If we apply to form also the optical lesson we thus learn with regard to color, it would seem that we ought to avoid square, monotonous masses, and regular, unbroken extent of surface in American rural architecture, because the climate rarely supplies the shifting, mellow light in which such simple forms appear to advantage.

The sky-line of a building should undoubtedly be determined, in a great measure, by the scenery in which it is to be located, and it may be either subdued or picturesque, according to the circumstances of each case. But the *plan*, which regulates the general design of the mass, and the *details*, may, with advantage, be picturesque in almost every situation; for, in this climate, the eye will be more likely to take pleasure in a rural composition that consists of a group of forms well connected and massed together into one individual whole, than in a study characterized by symmetrical uniformity, however complete it may be; for the former suggests, at the very outset, a freedom from effort, and offers the opportunity for a gradual examination, if preferred, whereas the latter must be grasped in all its completeness at once, and can only be truly enjoyed as a whole: and this naturally involves a more decided and continual effort on the part of the eye than is required in the other case. If the design is small, or on an economical scale, it may be inconvenient to have any breaks in the plan of the walls;

E

but some degree of picturesqueness can always be obtained by the treatment of the roof-lines, or by the use of verandas, porches, or bay-windows; and these features, if well arranged, are very valuable in any case, for they help to supply the variety of light and shade which is so much needed. The introduction of circular-headed windows, circular projections, or verandas, and of curved lines in the design of the roof, and in the details generally, will always have an easy, agreeable effect, if well managed; and curved roofs especially deserve to be introduced more frequently than has

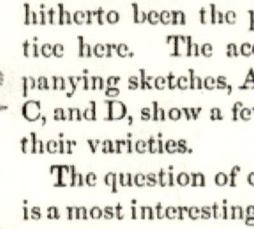

hitherto been the practice here. The accompanying sketches, A, B, C, and D, show a few of their varieties.

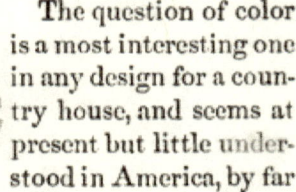

The question of color is a most interesting one in any design for a country house, and seems at present but little understood in America, by far the greater number of houses being simply painted white, with bright green blinds. By this means each residence is distinctly protruded from the surrounding scenery, and instead of grouping and harmonizing with it, asserts a right to carry on a separate business on its own account; and this lack of sympathy between the building and its surroundings is very disagreeable to an artistic eye. Even a harsh, vulgar outline may often pass without particular notice in a view of rural scenery, if the mass is quiet and harmonious in color; while a very tolerable composition may injure materially the views near it if it is painted white, the human eye being so constituted that it will be constantly held in bondage by this striking blot of crude light, and compelled to give it unwilling attention.

When a palace, like that of Versailles, is erected in the midst of formal gardens and terraces on a very large scale, and so arranged that it is the principal feature from every point of view, it is not inappropriate that it should be of white marble. There is nothing more interesting for the eye to rest upon than the building, and the light and shade of the architectural decorations, together with the general magnificence of the composition, are set forth to advantage; for pure white, even in large masses, is only disagreeable to the eye when it forces into prominent notice objects of secondary importance.

In country houses the design has to be adapted to the location, and not the location to the design, for it is undesirable, and generally impracticable, to make the natural landscape subservient to the architectural composition. Woods, fields, mountains, and rivers *will* be more important than the houses that are built among them; and every attempt to force individual buildings into prominent notice is an evidence either of a vulgar desire for notoriety at any sacrifice, or of an ill-educated eye and taste. The colors of rural buildings should be carefully varied. They should be often cheerful and light, sometimes neutral, seldom dark, and never black or white; and there is, fortunately, no end to the combinations of tints that may be used in painting a house. The constant recurrence of about the same requirements will, of course, lead to much similarity in plan, particularly in small buildings; but the monotony that this would occasion may be agreeably relieved by variety in color, both in the interior and exterior. Different patterns of paper will make two rooms of the same proportions no longer look alike, and the same result will be obtained on the exterior by adopting different tints for the walls and the wood-work. Another important point to be considered is, that it is entirely insufficient to use only one or two shades of color for each house. Every rural building requires four tints to make it a pleasant object in the way of color; and this variety costs but little more than monotonous repetition, while it adds much to the completeness of the effect. The main walls should be of some agreeable shade of color; the roof-trimmings, verandas, and other wood-work being either of a different color or of a different shade of the same color, so that a contrast, but not a harsh one, may be established. The third color, not widely different from the other wood-work, should be applied to the solid parts of the Venetian blinds, and the movable slats should be painted of the fourth tint. This last should be by far the darkest used on the premises, for the effect of a glass window or opening in a wall is always dark when seen from a distance; and if this natural fact is not remembered, and the shutters are painted the same color as the rest of the house, a blank, uninteresting effect will be produced, for when the blinds are closed, which is generally the case, the house, except to a person very near it, will appear to be without any windows at all. This error is often fallen into, and requires to be carefully guarded against. It is, however, a very simple and easy matter thus in a few words to lay down common-sense rules that may be advantageously followed in painting all country houses, but it is a very different affair to overcome the difficulties of ignorance and prejudice.

In some cases the house-painters themselves show a laudable desire to escape from monotonous repetition; but, on the other hand, they are often very troublesome opponents to reform in this matter. And this is not to be wondered at; for a mechanic who has been brought up on a chalk-white and spinach-green diet ever since he was old enough to handle a brush, can hardly help having but little taste for delicate variety, because a perpetual contemplation of white lead and verdigris is calculated to have the same effect on the eye that incessant tobacco-chewing has on the palate: in each case the organ is rendered incapable of nice appreciation.

Any person who may wish to have his residence judiciously painted will do well to depend on himself to make the selection of colors: and if he will but study the matter simply and fairly, trusting to his real natural instinctive taste, and will regulate his decision by his private feeling for what is agreeable or otherwise, instead of by what he finds next door to him, he will at once cut loose from conventional absurdity, and probably arrive at a result that will be artistic and pleasing.

It is highly satisfactory that, in this matter of color, which is so important to rural art, there is constant opportunity for improvement. The necessity for re-painting every two or three years fortunately compels the question to remain always an open one. Ill-planned roads and ugly houses are troublesome to alter; but improved taste may readily satisfy its craving for harmonious color, which will give, in every instance, a most liberal return for whatever outlay of thought or money may be judiciously bestowed on it.

After the plan and general design of the house are

decided on, and its prominent features are adapted to the peculiarities of the site, the point of construction yet remains to be settled, and it is one that affects both the comfort and the pocket of the party interested in a very decided manner.

Houses may be built of wood, brick, or stone, and each of these is well adapted for special purposes. Wood is undoubtedly an unsatisfactory material in one point of view, because it fails in expressing permanent durability; but there are many positions in which it may be used with advantage for out-buildings or small cottages, or for larger buildings that are required for temporary purposes.

Circumstances, indeed, may arise, in which it is desirable to construct quite good houses of wood. Thus, if the rough stone of the neighborhood is absorbent of moisture, the lime inferior, the brick porous, and good timber easily obtainable, a house built of wood, and filled in with common brick, will be the most comfortable and durable residence that can be erected without sending an inconvenient distance for material. It is, therefore, quite necessary to study the capabilities and varieties of wooden construction, although it is unquestionably inferior to a more solid and substantial method of building. One plan is to use vertical boarding, with battens to cover the joints, for an external covering. This mode has some advantages, and its appearance is often preferred. It is well suited for barns or small buildings, where the battens are relatively large enough to form part of the design; but when used on a larger scale, it is apt to give a striped, liny appearance to a house that injures its broad, general effect, and to draw particular attention to the fact that it is built of wood. This fact should not in any way be denied; but it is not desirable to make it especially prominent, as if it was something to be particularly proud of. It is for this reason that the ordinary mode of horizontal siding seems preferable in most situations. It offers a simple, fair surface, that can be broadly treated both in form and color, for the slight projection of one board over the other does not give sufficient variety of light and shade to interfere with the general effect as a whole. Another method is to groove and tongue the boards together, and bring all to one smooth surface. This plan has nothing to recommend it; it is more costly, more likely to get out of order by expansion and contraction, and is scarcely more agreeable in appearance. It is possible, instead of using siding, to cover a building with shingles, and to cut them into ornamental patterns. And this was often done by the Dutch settlers; but the projection is so slight, that not much additional effect is gained, except, perhaps, in quite small buildings, for the impression that a residence of tolerable size makes on the eye depends very little on such merely superficial detail.

There is a very picturesque mode adopted in Europe of building what are called half-timbered houses; that is, heavy frames of wood filled in with brick, and plastered on the outside on the same face as the framing-pieces. This plan is, I think, scarcely admissible here, except in particular localities, because the alternations of heat and cold are so great that, in time, the plaster is likely to separate from the wood, and the cracks thus formed must be fatal to the soundness of the building. In the neighborhood of Boston, however, I am told that there are some specimens of this mode of construction that seem to stand fairly.

One main thing that has to be attended to in wooden buildings is to make the corner-boards, the facias, the architraves, and base-boards *broad* and *heavy;* for unless this is done, the house, however strongly constructed, must inevitably *look* mean and contemptible. A double corner-board may be introduced in some cases with manifest advantage.

Pine is undoubtedly a better material than hemlock, yet the latter is much cheaper, and, if of fair quality, is nearly as good for constructive purposes as pine. It is, therefore, quite sufficient in all ordinary buildings to construct the frame, joists, partitions, and roofs of hemlock, using clear pine for the external and internal fittings and finish. Oak is the best, and, in the end, the most economical material to use for heavy timber across wide openings. Chestnut, in short lengths and well supported, is well suited for rough joists or sleepers required for boarded floors close to the earth; and locust-wood, though costly, is invaluable in moist situations for any posts, furring strips, or other wood-work that comes in direct connection with damp basement walls. For covering roofs of houses in the country there is scarcely any good material so generally available as shingles, if the pitch is not too flat. Slate forms an excellent covering, if of superior quality and well put on, so as not to be loosened or blown off in fierce storms. Tin expands and contracts, and has a tendency to get out of order, but still is a very good roofing material when properly put on. Zinc is worthless. Thick canvas is good for flat veranda roofs or small surfaces, being preferable to tin, inasmuch as it suffers less by alternations of temperature, reflects less heat, makes less noise in rainy weather, and takes less time to put on.

It has been the practice hitherto to depend on the Welsh slate almost entirely, and as this is but of one uniform gray tint, nothing is gained by its use as far as appearance is concerned. But lately, new American quarries, supplying slate of different colors, have been opened in various parts of the country, and worked with success. The slate that comes from the Eagle quarries in Vermont is of two tints: the one a rich purple-gray, the other a delicate green. This slate, when arranged on a roof in stripes or patterns, so that the colors are equally represented, has a very agreeable effect, and one that is far superior to that produced by any shingle or metal roof. Whenever slate is used, precautions should be taken to prevent any drift of fine snow under the slates. The joints should be laid in mortar, the boarding should be matched, and the pitch of the roof should not be at all flat. In some cases tarred paper is laid over the boarding as an additional safeguard from drift.

The great advantage of a shingle roof is, that it is scarcely possible for it to get out of order till the wood absolutely rots; and this takes many years to accomplish if the shingles are tolerable and the work well done. It also allows of considerable expansion, contraction, and even settlement, without the slightest injury to its efficiency. It is agreeably varied in surface, and assumes, by age, a soft, pleasant, neutral tint that harmonizes with any color that may be used in the building. A shingle roof in cities, or even closely-built villages, is objectionable in case of fire, because the loose, lighted chips from an adjacent burning building will be likely to inflame it. But this objection does not amount to much in a detached building in the country, the plain fact being that there is a risk; but, all things considered, it is worth while to incur it.

There are several methods of making wooden buildings capable of resisting the cold and heat, but none seem so good as filling in with common brick nearly the full thickness of the studs, and then lathing and plastering on the inside. This mode is simple, cheap, and, besides answering its more immediate purposes, it serves to keep out the rats and mice which are apt, after a time, to scramble noisily about a wooden house, or, what is worse, die and be disagreeable in some undiscoverable, out-of-the-way place behind the plastering.

Hard brick set in good mortar is an admirable material for building the walls of a country house, and is a mode that admits of considerable variety in construction and finish. An eight-inch solid wall may be used if the building is of moderate size, but it ought not to be weakened by building floor joists or furring timber into it. The wall should be a solid brick wall throughout. The floor joists should be supported on iron rests affixed to them, and built into the wall as the work proceeds. The furring strips should be the thickness of a mortar joint and half the width of a brick, so that, in the event of their decay, the walls will remain thoroughly sound. An attention to this important point in the construction of all walls is highly necessary.

In city architecture the joists are commonly built into the walls story after story, thereby materially weakening the brick-work, and causing a result in case of fire that is truly disastrous; for when a hole is burned in any of the floors, the unsupported joists, acting as powerful levers, very soon heave over the walls into which they are built, and then utter ruin of the building of course ensues.

In a range of stores, called Commercial Block, that we designed for Mr. Milton J. Stone, and which is now erected in Quincy granite on Commercial Wharf, Boston, the joists are supported on heavy stone corbels jutting out from the walls; and in a warehouse lately finished in Washington Street, New York, for Messrs. Robins and Co., the projections are of brick. The iron rests described above are thus rendered unnecessary; but in domestic buildings it is generally desirable to preserve an uninterrupted cornice line, and for this reason the iron rests, as they take up very little room, may be introduced with advantage in ordinary houses. One great point that is thus gained is to keep the timbers entirely clear of the damp external wall. Common sense can not fail to see the propriety of taking some precautions on this point; but if any other authority is asked for, we need not be at a loss, for biblical authority shows that it was considered in the construction of Solomon's Temple. In 1st Kings, c. vi., v. 6, we are told that he made narrowed rests round about, that the beams should not be fastened in the walls of the house.

Design No. 16, which is not a small house, is built in this manner, with iron corbels. The proprietor was perfectly willing that the brick-work should be one foot thick; but as the plan is so arranged that the walls assist each other, and are, moreover, strengthened with buttresses, the extra four inches of brick-work seemed a waste of material, and the house is accordingly built of eight-inch brick-work throughout, and appears to be, to all intents and purposes, a strong and substantial building.

Solid walls require to be furred off on the inside to receive lathing and plastering; and this applies as forcibly to a wall eight feet thick as to one that is eight inches. A wall one foot thick will be strong enough for almost any country house; and all the remarks that have been made with regard to eight-inch walls apply also to those that are thicker.

Hollow brick walls have many advantages. They are fire-proof, they keep out cold and heat very efficiently, they leave a good place for boxing or sliding-shutters, and are very firm and substantial if well constructed. A wall eight inches thick, with a hollow space of three, four, or five inches, and an inner wall of four inches, is the thinnest hollow wall that can be properly built, and it must be a very large building indeed to require any thing thicker. A course of slate laid in cement should be built in thoroughly the whole thickness of the wall on the line of the base-course, to prevent capillary attraction taking place between the foundation wall and the inner thickness of four-inch brick, on which it is proposed on each floor to plaster without lathing. The bond between the outer and inner wall should be of strips of iron, painted or tarred; for if brick bond is used, there will be a slight connection at intervals between the two walls, and in driving storms some damp may possibly get through. The two thicknesses of brick must be entirely and totally distinct if a satisfactory result is to be arrived at. In this trying climate one thing, also, must be borne in mind, that brick is a readier conductor of heat than wood, and, consequently, that a brick inner wall will absorb more of the heat of a room than a wall that is furred off. It will, however, retain the heat longer, and thus, when the house is once thoroughly warmed, it will appear to be the warmer mode of construction of the two. The fact, however, that the brick wall is so good a conductor of heat may operate prejudicially if the house is shut up and unwarmed for a lengthened period, because the moisture in the air being warmer than the wall will be apt to condense on its surface like frost on a window-pane, and the wall may possibly give signs of dampness, and even injure a delicate paper, when it is in reality as impervious to moisture as the frosted glass already referred to. This is an objection to hollow walls that it does not seem possible to overcome, and which has to be considered before deciding on their adoption in a house.

Many houses are built with hollow walls, and then furred off on the inside; but there is more labor in constructing a hollow wall than a solid one, and the advantage is not very apparent when furring is used, except sliding-shutters are required.

The outside surfaces of brick walls may either be left in their natural state, or covered with a coat of boiled oil, or with a lime-wash, or painted or cemented either with or without a subsequent painting. When the materials are good and the color not over red, ordinary brick-work in its natural state, if pointed with dark mortar, has a very good effect for country houses, and it might with advantage be more generally left untouched than it is. Creepers would then grow on it without being interfered with, and the annoyance and expense of lime-washing or painting every year or two would be avoided. A coat of boiled oil has a good effect where more expensive face-brick are used, both in hardening the surface and equalizing the tint. A lime-wash in two coats is a cheap and effective mode of covering brick-work, and several agreeable tints may thus be obtained. Painting a wall three or four coats of good oil-paint is a better, though, of course, more expensive method. The lime-wash sometimes acquires a mouldy look on the stormy side of the house. This is avoided by the use of paint, and any shade of color can thus be arrived at. Cement may be used with advantage when the quality of the brick is poor; but it is undesirable to conceal a good brick wall which is a genuine, solid, and permanent piece of construction, under a coat of cement, which offers a less durable surface. It is not often, indeed, that cement stands perfectly in this climate. It is liable to scale off after a year or two; and although numbers of instances might be cited in which it has been quite successful, covered with a projecting roof and thoroughly painted, it is in no way, that I can see, superior to a good, hard, well-constructed brick wall. A cement surface may, indeed, at a considerable expense, be lined, tinted, sanded, and scored up and down till it becomes a tolerably fair portrait of free-stone for a few months; but although some persons may consider the result handsome and satisfactory, it will not fail to convey to others the same disagreeable impression that is produced by seeing the human countenance dressed up with rouge and false hair.

For covering walls of porous brick or inferior stone cement is available enough; and those who are constantly engaged in its manufacture and use speak highly of it as a material for general external work, even without painting. Still, I should always prefer a good brick wall. Yet circumstances will arise in which it is desirable to try the experiment of cementing; and I therefore transcribe a portion of an interesting letter I lately received from Mr. Tompkins of the Rondout cement quarries on the subject. There can, indeed, be no doubt but that the success of the work depends quite as much on the way in which it is done as on the material used.

"First, saturate the surface to be operated upon with water used abundantly (a force-pump and hose is the best method). Secondly, make a wash of liquid cement, as for inside brick walls, applying it with a brush, so that all small cavities may be entirely filled. Then spread on the finishing coat about a quarter of an inch in thickness, and made in proportions of two of sand to one of cement. During the operation of putting on this second coat, the first coat of liquid cement should be kept quite damp by frequent sprinkling. After the cement is upon the wall, it is important that it should be sprinkled with water, so as to keep it damp for a week or two.

"In making the mortar care should be taken to have none but clean, sharp sand, free from loam and quick-sand. If taken from the sea-shore it should be very thoroughly washed with fresh water before using. All the cement required is just so much as will be sufficient to coat each particle of sand. The sand and cement should be thoroughly mixed before water is applied, and water should be applied to only so much as will be used immediately. The above method was adopted in stuccoing the walls of a house about ten years since, and they are now as perfect as when first coated."

When thoroughly trustworthy cement can be found that does not require painting, some agreeable and honest effects may be produced by stamping it with a pattern; but till then it is a material that had better be used but sparingly. In England the use of cement for surface-work has exercised a most pernicious effect on architecture. "Compo" fronts now swarm in town and country. Cheap materials cheaply run up, and smeared over with Portland or Roman cement, impose for a year or two, perhaps, on the passer-by. But the day of reckoning soon comes, and the peeling, cracking, spotty complexion that the false stone assumes proclaims the whole affair a worthless sham and contemptible failure. Fortunately the use of external cement has made no great progress in America at present, and it is to be hoped that the more genuine modes of construction will continue to be preferred.

A very agreeable and superior method for rural buildings is to combine stone of some quiet color, like brown stone, with brick of any color, oiling the brick once for all, but leaving the stone intact, as one coat of oil will utterly destroy its texture and appearance. Wrought stone being an expensive material, this method is not likely to be adopted except in large houses; but it is well worth while to consider that a great ultimate saving may be sometimes made by a comparatively small increased outlay in the first instance, and it certainly does seem unsatisfactory in a house of any pretension that it should every now and then require to be covered all over with some superficial coating to make it agreeable.

The simplest method of building with stone is to use the gravel wall. This is a wall made of concrete, and is a method of construction of considerable antiquity, and common in various parts of Europe. It is described in detail in Mr. Fowler's work on the subject. Under favorable circumstances it may be cheaper than brick; and as it makes as good a wall when thoroughly executed, there are many situations in which it must be desirable to use it. But it is not much in demand at present, and has no perceptible advantage over a rough brick wall. It, moreover, requires to be plastered or cemented on the outside surface, which is troublesome.

Rough stone walls for cottages and country houses have a good effect if the situation and accessories are consistent: and a very sound and picturesque-looking wall may be constructed by using small, rough stones for the body of the work, and red brick for the angles and window-dressings. It is uncommon, however, in this country to find a taste suited to this bold, unconventional treatment, and undoubtedly for a year or two, till the creepers have grown up, it may look rough; but the ultimate result is most agreeable and country-like, and such a surface will require no painting or lime-washing, and will mellow and improve every year.

Houses that are built of squared brown stone have a melancholy, dingy, monotonous, and uninteresting look, and it is a material that ought not to be used in large, flat masses. Marble is too white to be agreeable in the country. Squared blue stone is cold, prison-like, and repellant for ten or twelve years, but it then begins to be affected by the weather, and, by degrees, assumes a very beautiful, soft, golden-gray tint. Ten or twelve years, however, is a long time to wait in America. Granite is still colder and more expressionless than blue stone, with the additional disadvantage that it is wholly uninfluenced by time. Stone from Caen, in Normandy, has lately been introduced into New York, and is used to some extent. It is a beautiful material, and very delicate in color, but unequal in quality, unless *specially imported from well-*

F

known firms. It seems, however, a little unnatural for a continent like this to seek building materials in Europe, and there can be little doubt but that a strict geological examination will, after a time, supply us with many new varieties of building-stone. A capital free-stone, of a pleasant, soft tint, has lately come into use, brought from the Dorchester quarries, Westmoreland County, New Brunswick.

As it is only proposed in these remarks to give a cursory view of the more prominent building materials and modes of construction, it is scarcely necessary to pursue this part of our subject into all its various branches, and we may take leave of it with the general recommendation to those who are about to build, to spare no pains to obtain the services of honest, intelligent master-mechanics, as deficiencies of construction and execution, and heavy bills of extras, are more frequently the result of dull incapacity and stupid neglect than of a grasping disposition, or of a willful intention to deceive.

It only remains now, before entering upon the general description of the designs that form the staple of this volume, to take a passing glance at the principal features of detail that occur in country residences; and the *porch*, or *entrance*, suggests itself as having the priority of claim to our notice in this respect. This part of the design is the first that appeals to the attention of a visitor, and admits of much character and expression. A large house may be spoiled by a mean porch, and the interior effect of a small, compact house will appear dwarfed and contracted if it is approached through a pretentious entrance-porch. Experience has developed several marked varieties of design in this prominent feature of every attractive rural home, and each has its distinctive peculiarity of plan or purpose. The simplest and most economical way to obtain the temporary shelter from heat, cold, or storm, that it is the office of a porch to provide, is to shut off a recess from the interior of the hall, or passage-way, by an in-

ner door. This method is shown at A, and answers in practice fairly enough, but it does not *express* its purpose of shelter very decidedly. It suggests the idea of an easy, convenient entrance to the interior of the house, but, in many situations, we want more than this. We need something that shall indicate the protection from the weather that the porch offers, and give life, and light, and shade to the design. This result may, in a measure, be obtained by putting a boldly-projecting hood over the opening; and an instance of this method is illustrated at B. The sketch of an elevation given at

C may serve to explain a variety of porch design that deserves much more attention in this climate and country than it has yet received. It is intended to be constructed of timber, framed and braced, and may be either put together in a rough, rude way, and covered with vines and creepers, or elaborated with any amount of decoration that may be considered suitable. It should project  eight or ten feet from the building, even in small cottages, and have a bold, overhanging roof. This style of porch almost supplies the place of a veranda on the side of the house where it occurs, and may be fitted with glazed frames and an outer door for winter use, if preferred. The porch shown at D is also of wood, but in a somewhat different style. In this case it is proposed to supply an inclosed vestibule at all seasons of the year, but the outer doors would probably be unhinged and put away during the summer months. Sometimes a veranda, such as E,

may supply the place of a porch, and serve as a protection to the front entrance. It will not probably be convenient to inclose entirely a veranda-porch at any season of the year, but the open *end*, or ends, can easily be shut up during the cold months; and if this precaution is taken, it will make a considerable difference in the value of the shelter afforded in stormy weather. It is not an uncommon practice to provide a little plain box-porch, the size of the front door, for  winter use under verandas; but this arrangement is sure to be unsightly, and to have a meagre, incomplete look. Where a veranda runs the whole length of a front, and the porch occurs in connection with it, it may with propriety be either of brick or stone. F shows a simple arrangement for brick, and G a more elaborate study for stone construction.  Another method is to inclose a projection, and put a balcony over the entrance-arch. This plan, as will readily be seen, admits of considerable variety of design, and its principle may be understood by means of sketch H.  The most complete arrangement is to project the porch clear of the building, and arrange a carriage-way through it. But this plan is scarcely suitable to any but large, handsome residences, not alone on account of its cost, but because the minimum size actually required for the convenient admission of a carriage is not likely to be in good proportion with the other parts of a villa of comparatively small dimensions.

An example of a *porte cochère* of this sort, with a single arch on each side, is shown at I; and another, with three openings in front, and one elliptical arch at the side, is shown at K.

Porches may be fitted up with permanent or movable seats; but these should be of wood in preference to iron, as the latter is a very cold, unyielding, and unsatisfactory material for such a purpose. They may be floored with wood, or paved with stone or marble, in two tints, or with a

simple pattern of ornamental tiles. The latter are not expensive, and have as cheerful and pleasant an appearance as any thing that can be used; but they need, in laying, much more care than is usually bestowed by country masons, as they are liable to get loose and crack with changes of temperature, if the surface is at all irregular or uneven. The pattern should be very simple, as, otherwise, an expectation of richness and elaboration in detail may be raised on the threshold, to which the interior of the house may fail to respond, and a disagreeable inconsistency will thus be apparent, which the remembered beauty of the paving will in no degree help to reduce.

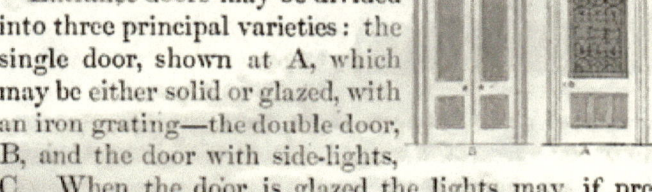

Entrance-doors may be divided into three principal varieties: the single door, shown at A, which may be either solid or glazed, with an iron grating—the double door, B, and the door with side-lights, C. When the door is glazed the lights may, if preferred, be hung on hinges from a centre rail, so as to let air into the hall without opening the door. The side-lights should always be hung as sashes. In some cases a sliding or folding inner solid door is construct-

ed for protection, but the iron grating answers all the purpose, and the glass may be ornamental, so as to give the light required without exposing any view into the hall from the porch; and it is always worth remembering that a glazed door illuminated from the inside has at night a much more agreeable and sociable appearance to any one approaching than a solid hall-door.

The porch leads into the *principal hall*, which should connect easily, and, to some extent, symmetrically, with the rooms. It materially lowers the character of a hall, especially of a large one, if the positions of the various openings, etc., are unstudied and irregular; two doors, perhaps, clustering together at one spot, while a third lonely one is penned up tight in a corner, with insufficient room for the casing round it. The result in such cases is an appearance of *carelessness*, not of *freedom*, in design. A shows a sketch for the end of a hall of but moderate dimensions, in which are many doors. By this plan two doors are grouped together, with a panel and solid bracketed shelf between them, thus avoiding the monotonous effect of a series of doors of the same height,

A

the sides being treated in a similar manner, and filled in with pictures in the panels. B shows the side of a larger hall, in which the blank space that

B

would otherwise occur between the two doors is fitted with an arched recess and seat on one side, and with console-table, looking-glass, hat-pegs, and clock on the other. C illustrates an open hall, with a staircase beyond. In this plan the upper flight of the principal stairs is supported on the arcade, and the two halls being thus connected together, a light, airy effect is produced, that is free from any practical objection,

C

if a servants' staircase is also provided for on the plan.

D illustrates a method of getting over the difficulty of an opening at the extreme end of the side of a hall. A light screen marks the passageway, and gives regularity and some individuality to the design without making it less light or airy. The plan of a hall in which this arrangement is introduced may be seen on the principal floor of Design No. 12. E is a similar example, but of a less expensive and more open character.

From the hall we may proceed to the *library*, which is an apartment in very general demand in most country houses. It need not be of large dimensions under any circumstances, but should be so arranged that, even when occupied by only one or two persons, it may have a cheerful, domestic look. It may be adapted to its special purpose in various ways. The simplest mode is, perhaps, to recess book-cases,

as shown at O, on each side of a door, fitting each with a dwarf closet for papers or magazines, and these recesses should be prepared for while the house is being constructed.

P shows another somewhat similar arrangement, but  with glazed doors and without the dwarf closets underneath. In a library the book-cases fill it up considerably, and sometimes a door of communication is needed where it will occupy valuable space and be unsightly. This difficulty may be overcome by having one of the book-cases hung on hinges, and made to answer the purpose of a door; or an imitation book-case, with book-backs corresponding  with the other volumes, may be used. Q illustrates an example that has been used in a small, symmetrically-planned library, the only visible approach to which is from the principal hall by sliding-doors; but two of the book-cases are hung and fitted with an inconspicuous catch, so as to swing when needed, books and all, and a private communication is thus afforded with a boudoir on one side, and with the staircase-hall on the other.

In a really large library the books should occupy the whole wall space, except a range of wide, inclosed shelves all round for unbound books, etc. If an extensive library is needed to a house of but moderate dimensions, it seems desirable to divide it by projections into compartments, so that its height may not appear out of proportion to its length and width. Such an arrangement is shown at plan R. The library proper is thus divided, so far as its appearance to the eye is concerned, into four separate parts; but it is nevertheless, to all intents and purposes, practically one large

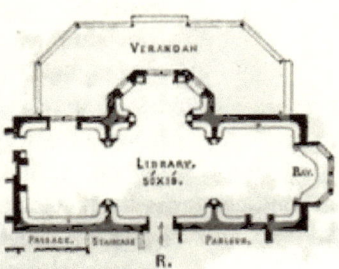

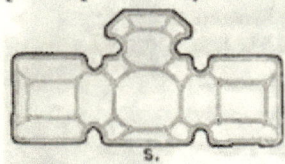

room. Plan S shows the design of the ornamental, ribbed ceiling, which is so arranged that there are many transverse lines, and but few in the direction of the length of the room, the eye being thus drawn to the separate compartments as much as possible. T shows an elevation of the end of this room, which has lately been erected, in connection with other improvements, for Judge Kent, at Fishkill Landing, and now contains his valuable library

of eleven or twelve thousand volumes. The general appearance of the exterior of this library, with the rooms over it, as finished, is shown at U.*

---

* This study may also serve as a further illustration of the remarks made with reference to alterations and additions in describing Design 17. The house, as purchased, was a fairly-proportioned, square country residence, with a kitchen wing so injudiciously arranged that it obscured the most attractive view from the windows of the principal rooms. The existing wing building was, therefore, taken down entirely, and the new library was designed in its place,

V illustrates a design, made for my own use, for a small book-case that can be screwed to the side of a room. The doors are glazed, and the whole affair takes up but little space, and is suited to a small library or study, as it allows of a table to be set against the wall beneath it. Such a book-case, made of black walnut at a cost of $25, holds about eighty or ninety ordinary volumes. W shows a sketch of a somewhat larger movable book-case, planned by Mr. Withers, for a recess beside his fire-place. The upper central space is uninclosed, and filled with shelves for the reception of small objects of interest. The compartments on each side are glazed for books. The central compartment below is for bound volumes of journals. The side spaces are fitted up with shelves and doors as lock-up closets for papers, etc. This book-case is made of oak, with the introduction of black walnut

so that a wide veranda now commands the extensive prospect that before was lost, and the kitchen offices were rebuilt to the north of the library, so as to shield as much as possible this unusually large room from the winter storm. The staircase was planned anew, and an inclosed veranda (so arranged as to protect the otherwise exposed northwest corner of the library from the cold) supplied large pantries and other useful accessories on the principal floor. In other respects the old house was left intact, except that a bath-room and water-closet were provided for in the chamber plan, and proper provision was made for warming and ventilating. In the new building several other bedrooms were provided over the library, and the house, at an additional cost of about $11,000, was made as complete in every way as the circumstances of the case admitted. The result, I am led to think, was satisfactory, as Judge Kent, in a note of the 12th January, 1855, in which he speaks of the great advantage derived by his servants in extremely cold weather from boarding up temporarily the veranda to the north of the kitchen wing, adds, "I beg to say that, on my part, I am much pleased with the result of your labors. My house is very much to my taste. The exterior is, I believe, generally admired, and the interior arrangements make it the most comfortable house I ever lived in."

for the mouldings, and cost $55. X is a little sketch for a hanging book-shelf, suited to a bed-room or small study. Such designs, however, come, perhaps, more under the head of furniture than architectural arrangement.

In the *dining-room* several useful features may be architecturally managed. The side-board, for example, may be arranged in a recess, as shown at A, with a door to a private closet on one side, and a pantry, or service-room, on the other. This idea admits of numberless modifications, and has always a rich effect if well managed. At B is illustrated the end of a dining-room opening on to a plant cabinet, or small conservatory. The glass doors of communication are fitted with slightly-ornamented glass, so as to decrease the monotony of effect that would otherwise occur, while enough clear glass is left to give a good view of the flowers, etc., when the sliding-doors are closed. And it may be worth while to remark here, that such a conservatory should be arranged for the display of flowers rather than for their

cultivation, which may be more conveniently attended to elsewhere. C is an elevation of the side of a dining-room designed and executed for Mr. H. W. Sargent. The three woods, black walnut, oak, and yellow pine, are used in combination, and the result, in execution, shows how valuable these woods are for internal decoration when used in their pristine simplicity and merely oiled. This effect of color, which is a principal part of the design, is not, of course, to be understood from the annexed sketch, which is a mere indication of the general idea. In this room the left hand panel slides up easily, and discloses a large dumb waiter, or lift, communicating at once with the cooking kitchen. On the opposite corner of the room is a self-balanced trap-door in the floor, which supplies, by an easy staircase, the only means of access to a small, private brick cellar, properly ventilated. This door can be lifted up, by a little key fitting into a patent lock, at any time, the carpet being suitably arranged, and the cellar is consequently under complete and convenient control by the master of the house at any moment. The butler's pantry is close to the dining-room, but a passage intervenes, which connects with a lavatory and water-closet, and also with a garden and conservatory entrance. These latter are all close to the dining-room, without connecting directly with it. The wall panels are filled in with a richly-embroidered material, and

c

the ceiling is paneled in plaster, and tinted, to harmonize with the rest of the design. Permanent sideboards, introducing the three woods, and suitable in style to the room, are put up; and the chimney-piece has also been specially designed for this apartment. It is illustrated as a vignette to Design No. 19.

The pleasant and harmonious result that may always be obtained by a judicious treatment of unpainted materials in the interior of a house is well deserving of consideration. Ceilings and floors may be decorated in this manner as well as side-walls; and D may serve as

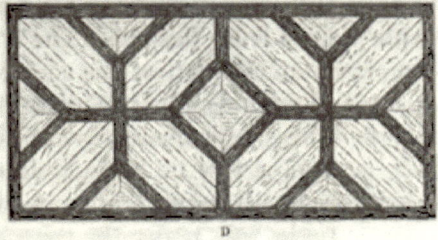
D

a hint for a simple floor in two tints, adapted to an entrance-hall or small study.

The *drawing-room*, or best parlor, next suggests itself to our notice. This room, although intended to be a strong point in every American house, is often made its least satisfactory feature. I have noticed one style, for example, which, in all probability, most of my readers have also seen. The walls are hard-finished white, the wood-work is white, and a white marble mantle-piece is fitted over a fire-place which is never used, as there is a stove in the room or a furnace in the house. The floor is covered with a carpet of excellent quality, and of a large and decidedly sprawling pattern, made up of scrolls and flowers in gay and vivid colors. A round table with a cloth on it, and a thin layer of books, in smart bindings, occupies the centre of the room, and furnishes about accommodation enough for one rather small person to sit and write a note at. A gilt mirror finds a place between the windows. A sofa, by courtesy so called, occupies irrevocably a well-defined space against the wall, but it is just too short to lie down on, and too high and slippery, with its spring, convex seat, to sit on with any comfort. It is also cleverly managed that points or knobs (of course ornamental and French polished) shall occur at all those places toward which a weary head would naturally tend, if leaning back to snatch a few moments' repose from fatigue. The sofa is, indeed, the "representative" man of the room, and concentrates in itself the whole spirit of discomfort that reigns unmolested in every square foot of the apartment. There is, also, a row of black walnut chairs, with horse-hair seats, all ranged against the white wall. A console table, too, under the mirror, if I remember rightly, with a white marble top and thin gilt brackets. I think there is a piano. There is, certainly, a triangular stand for knicknacks, china, etc., and this, with some chimney ornaments, completes the furniture, which is all arranged according to stiff, immutable law. The windows and Venetian blinds are tightly closed, the door is tightly shut, and the best room, that I am now thinking of, is, in consequence, always ready for—what? for daily use? Oh, no; it is in every way too good for that. For weekly use? No, not even for that—but for *company* use; and thus the choice room, with the pretty view, is sacrificed, to keep up a conventional show of finery that pleases no one, and is a great, though unacknowledged, bore to the proprietors. Such is one style of best parlor to be found in America; and though it is by no means universal, it is far too general for comfort. A drawing-room like this becomes a sort of quarantine in which to put each plague of a visitor that calls; and one almost expects to see the lady of the house walk in with a bottle of camphor in her hand, to prevent infection, she seems to have such a fear that any one should step within the bounds of her real every-day home life. All this is absurd. No room in any house, except, perhaps, in a very large mansion, ought to be set apart for company use only. If a reception-room for strangers is needed, it should be a small, unpretending room, certainly not the most agreeably situated apartment in the house, *which should be enjoyed daily*, for it is not the having any good thing, but the using it, that gives it its value. A friend of mine, when making arrangements to rent a small suburban house, happened to remark that he should occupy the back parlor as a dining-room, and the landlady seemed really quite overwhelmed with the idea, which she evidently thought an unwarrantable innovation. "All the best families," she said, "lived in the basement. Why use such a beautiful parlor merely for an *eating* room?" as if eating was a degrading occupation. Let us return, however, to our drawing-room. A best parlor ought to express, in its proportions, colors, and arrangement of furniture, an agreeable, hearty, social welcome. The lady who studied her room when her guests had departed, after a lengthened and agreeable visit, so as to learn how the furniture had accommodated itself, as it were, to suit the social convenience of her friends, and who then modified her previous ideas accordingly, had the true artistic eye for beauty of arrangement, and certainly deserved to have a pleasant circle of acquaint-

G

ance. There are but few strictly architectural features in a drawing-room that call for illustration. Good proportions can be supplied; but the lady of the house is the most important architect here. A bay-window is a very desirable addition to such a room, as it breaks up the monotony of outline, and gives a more free and open effect. Studies for interior arches to such windows are shown at A, B, C, and D.

As we have now passed through the principal rooms, we may proceed to take a glance at the chamber plan. The *staircase* is a very characteristic feature in a house, and a convenient adaptation of the height of risers, breadth of treads, height of railings, plan of landings, and finish of newel posts, adds more than would, at first sight, be supposed, to the general comfort of the house; for this part of the design is common to all of the residents, and is, moreover, sure to be in constant use. A simple and economical arrangement is to have a plain turned newel post and turned baluster. This

plan also admits of much ornamentation, and an enriched specimen is shown at L. M illustrates an oc-

tagonal newel post. N shows a staircase of higher character for a large hall, with continuous string and newel post at each landing; and such a staircase has a fine effect when well constructed of hard wood, particularly if it is fitted with a skirting, or dado, and half hand-rail against the wall, following the rise of the stairs on the same line as the main hand-rail. O shows a design of still higher pretension, in which the balusters are dispensed with, and the space filled with open tracery. In this idea of a design all the forms are made to adapt themselves to the general upward line  of the staircase, and a more easy and graceful effect can be arrived at in this way than in any other. It involves, however, like most superior arrangements, an increase in cost, and can, therefore, only be adopted occasionally in large houses.

On arriving at the chamber floor we naturally expect to find a *bath-room* and *water-closet*. Some persons have a prejudice against planning these two conveniences in the same apartment; but my own experience is in favor of placing them together in a house of moderate size, as the advantages of saving of space, free ventilation, and plenty of room to move about in, more than compensate, in practice, for the slight inconvenience that must certainly occur now and then, if both bath and closet happen to be required for use by different members of the family at the same time. It may be well here to advise any one who fits up a water-closet inside a house, to ventilate the *basin* with a half-inch pipe carried behind the furring into a flue, as well as to ventilate the *room* by a register near the ceiling. A house-maid's sink saves a good deal of trouble if properly managed; but it should not be in the bath-room, but near the attic stairs, or wherever it can be most privately located, so as to be well lighted and available for constant use. A speaking-tube, with whistle attached, should be fitted to connect this floor with the kitchen department; and a dust-shaft, if well placed and judiciously planned, is a labor-saving addition to the accessories, as is also a lift for coal-hods, or to send down clothes for the wash, if it can be cleverly managed so as not to waste valuable space. A linen-press is a universal requirement; but, with the exception of a roomy closet for towels, etc., this convenience, in a moderately-sized house, may just as well be provided for in the attic, where there is likely to be more room to spare. Any superior room or press for woolens or linens should be supplied with cedar shelves.

We have now fairly arrived at the *bedrooms*. A bedroom for a lady's use, to be complete, should have two windows on one side, so that a looking-glass and dressing-table may be placed between the two lights, and, if possible, a window on one of the other sides, so as to obtain a thorough draught through the room whenever it may be required. Each bedroom should have a ventilator in it, and one or two closets. These may be arranged in connection with the windows on the sides or ends of the room, if preferred, so as to improve the  appearance of the apartment, and give a recessed bay and seat. Q and R show two studies for this sort of arrangement, which, as will be seen in the designs to be hereafter described, I take every fair opportunity of introducing.

Attic bedrooms may be so planned as to afford a valuable addition to the accommodation of a country house. A more extended view is generally to be obtained from these rooms than from the rest of the house, and as they may be made quite as comfortable, though not, of course, so symmetrical, as the second floor chambers, they deserve a fair share of consideration; and the economical advantages they offer have frequently led me, in practice, to advise the use of a high-pitched roof, with a flat on the top. The acute angle of the roof precludes the possibility of any large surface being exposed to the vertical rays of the sun, and the flat on the top, being furred down some three or four feet, supplies an air-chamber above the attic ceiling that acts as a satisfactory shield from the heat. This space can be floored and fitted with a staircase, and used as a lumber garret, if thought worth while. This plan of arranging the roofs, so as to increase the value of the attic rooms without loss of exterior effect, has been so generally approved, that I can, with confidence, recommend it to the examination of all who intend to build country houses of moderate size. The flat is not noticeable from below, as the ridge lines alone are apparent; and many to whom I have explained the principle of arrangement by an actual visit to executed houses, have expressed their surprise at finding a large, nearly level space, on the top of a house that showed no sign of any thing of the sort to a passer-by. Some further remarks are made on this subject in describing Design No. 7. Studies for arranging attic windows and closets are shown at G and H.

The *roofs* and *gables* of a country house may be designed in many different ways, and some of their principal varieties may be seen on reference to the accom-

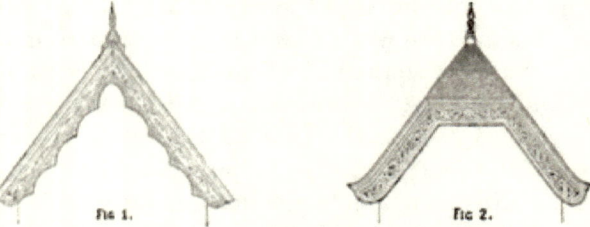

panying little sketches. Fig. 1 is a high-pitched gable, with verge-board. Fig. 2 shows a similar gable hipped back, which entirely alters its character. The eaves may be curved, as here shown, if preferred. Fig. 3

shows a gable of flatter pitch, with cantilevers. Fig. 4, a corresponding gable, with pendant finish. Fig. 5, a

curved gable, with pendant finish. Fig. 6, a roof with a single convex curve. Fig. 7, a concave curve. Fig.

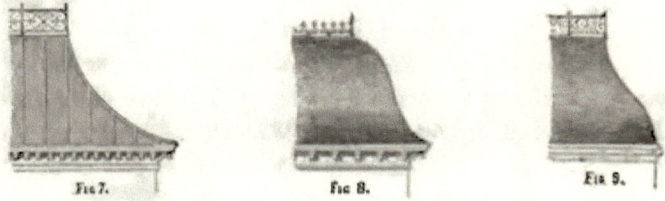

8, an ogee curve; and Fig. 9 another form of ogee curve. Each of these forms is available in one situation or another, and I believe all are introduced in the accompanying designs.

Dormer-windows are of several sorts, according to the style of the house. They are often made *too small*, and considerable comfort and effect is thereby lost, for a small one costs very nearly as much as a large one, and is not half so available. A dormer is a capital feature in a country house, and never need be ashamed of itself, or try to shrink out of sight.

A shows a study for a single large window. B, another one, hipped back, with curved eaves. C, a

study for a dormer with double window; and D a picturesque arrangement for bringing a dormer forward, and making it a more important part of the design. The perspective effect of this latter arrangement may be seen in Designs No. 4 and 31. A dormer in a different style is shown as a vignette to Design No. 22.

Verge-boards and *finials* admit of endless variety of design, and some studies for these details are shown at E and F, and in the vignette to Design No. 21.

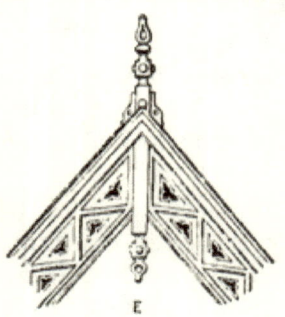 

Rain-water pipes, as generally planned, are most unsightly accessories to a country house. They need never be so. On the contrary, they may often be made valuable helps in design, as they can be used to mark a vertical line in the composition where there is no projection in the plan. Instead of being circular, as generally made, they should be semi-circular, so that they may rest flat against the wall surface; and they should also be fitted with heads, and the pipe that conveys the water from the roof should be carried into them with a curved line from the level of the gutter, instead of being cut through the cornice in the common, slovenly, broken-backed way that disfigures nineteen out of twenty country houses in Amer-

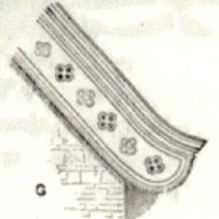 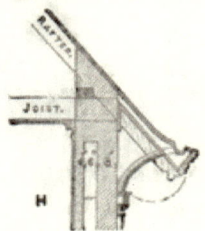

ica. G and H are illustrations showing the foot of verge-board and construction of gutter in a roof with curved eaves. The mode of continuing the gutter to the rain-water head applies also to any roof in which the gutter is cut out of the rafters. Some designs for heads are shown at I and K.

Chimneys are the very first things that catch the eye, and the last to escape from observation. They should, therefore, be well studied, and have a substantial, hospitable look. All clay *pots*, pretty or ugly, should be

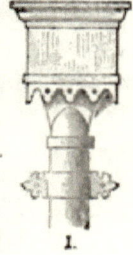 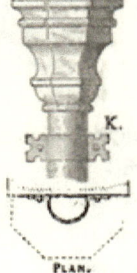

faithfully eschewed, for they are the most mean, shabby, dwarfish features that can be added to a house. A plain brick shaft, without any relief whatever, is to be preferred to the most fanciful *pot* that money can purchase. But there is no need to make the brick chimney so very plain; the flues may be grouped together in many different ways, and set-offs may easily be made in the brick-work. A shows a stack of six flues, simply finished with a blue stone on the top. Chimneys should always seem to stride the ridge, and never appear to sit on it side-saddle fashion, for a very disagreeable monotony of line will be the result if this point is not attended to. Where a case of this sort occurs, it is

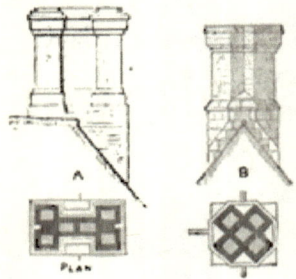

best to work out to a square or cross before leaving the roof, such as B. A plan is shown as a vignette to Design 24, in which, as there are many flues introduced, the ventilating openings are arranged at a lower level, to reduce the heaviness of effect that would otherwise ensue. A stone and brick chimney-stack is shown at C, and an elaborate stone shaft at D. As a general rule, it is preferable to plan the interior of a house so that the chimneys may start from the ridge, as it saves heat and

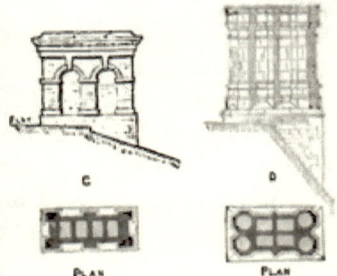

prevents any chance of the chimneys smoking. But circumstances alter cases, and it is now and then almost necessary to plan them in the outer walls. A few specimens of this style of chimney may be seen in the perspective views hereafter submitted, such as in Designs No. 3 and 37.

Ventilators are often useful both for convenience and artistic effect. Numbers 1, 2, 3, and 4 show some

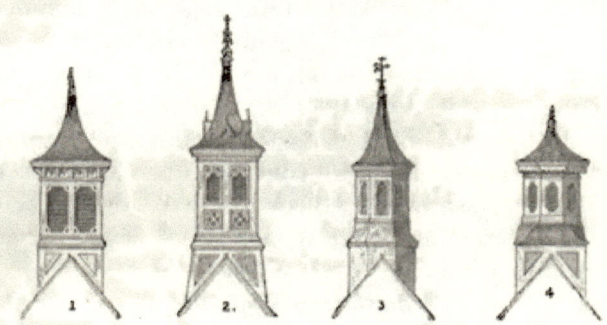

of their varieties of design that have occurred to me in practice. They can conceal an Emerson's ventilator, if preferred; but if the air-pipes are brought together at this point near the ridge of the roof, the simple outlet will, in most situations, be found sufficient. They need to be planned with a proper escape for the water that will find its way into them in rainy weather, or the practical result will be unsatisfactory.

Hoods to windows in American country houses are features that seem to spring naturally from the peculiarities of the climate, and the needs they give rise to. The upper sashes of windows with hoods can always be left a little open without any chance of the rain beating in; and even when of small size they protect the glass from the direct vertical rays of the summer sun, and receive the first blows from the winter storm. They also add much to the artistic effect of a rural building, and deserve, I think, a more full appreciation of their merits than has hitherto been accorded them. A few examples are shown at A and B, and as vignettes to Design No. 30. It is to be observed that, in summer, a small window is in one

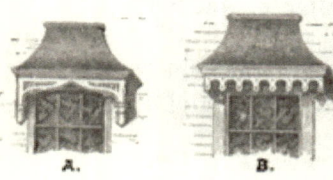

respect most comfortable, as a wall is a better protection from heat than glass or Venetian blinds, and most of the Italian and Eastern villas are planned with small openings. But, on the other hand, large windows are desirable to throw open for the summer evening breeze, and to let in plenty of cheerful light during dull winter and spring days. The hood, in a measure, connects these two opposite needs. A veranda all round a house is delightful for a month or two in the heat of summer; but most healthily-constituted persons like to have the opportunity to admit a stream of glorious, warm, genial sunlight into their rooms whenever they feel inclined to enjoy it, and this can not be obtained if the veranda entirely encircles the living apartments. The hood, on the other hand, defends the window from the powerful rays of the mid-day sun without shutting it out entirely.

The *balcony* is a feature that can now and then be introduced to advantage; and a specimen that tells the whole story, and scarcely needs any further detail, is shown in the illustration to Design No. 24.

The *bay-window* is the peculiar feature next to the veranda that an American rural home loves to indulge in. There can, indeed, scarcely be too many for the comfort of the house, or too few for the comfort of the purse, for I regret to add that they are expensive features. The simplest form is a plain semi-octagon.

with simple shed-roof, shown at L. This sort of bay is very commonly finished with a roof running up to a point against the wall; but the effect thus produced is always mean and disagreeable, and a straight line for some distance, as shown on the sketch, gives the appearance of the windows *belonging* to the house much more  than the other mode. M shows a plain bay with a balcony over, so that the occupant of the room above may be able to step out and see the view, or attend to a few choice plants. The square bay shown at N gives much more additional room than the octagon, and if connected with a continuous balcony and arch beneath, as there shown, has a handsome architectural effect; but the

other form is generally to be preferred. Circular bays are effective, but more expensive than any others. In the design of a bay-window a great deal will always depend on the arrangements made for the shutters or blinds. A want of forethought often leads to the allotment of so scanty a space for the piers at the angles, that the shutters are found to be in the way, except when closed, which is, of course, annoying. A large bay window, with smaller bay above, is shown at O; and P shows a bay with a covered balcony over

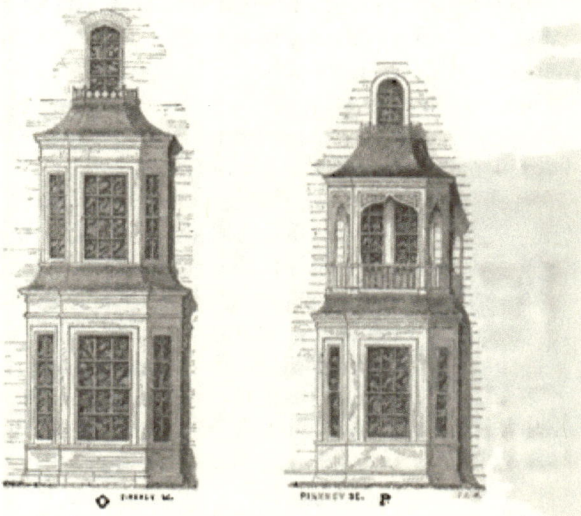

it; and the effect of this combination, which admits of much variety of design, is, in execution, very airy and elegant, if properly designed and solidly constructed.

Recessed arches and *arcades* will always produce a capital contrast of light and shade if introduced with any judgment, and deserve, I think, more attention than they seem to receive. They differ materially, both in accommodation and artistic effect, from verandas. The arcade being inclosed on three sides, affords much more shelter from the weather, throws a deeper shadow, and is more secluded from observation; while the piazza, which is always exposed on two sides, and generally on three, is more open to the cool breezes in hot weather. Arcades should, therefore, be introduced in connection with, and not instead of verandas. R and S show two varieties, which, however, can scarcely be judged of in mere outline. Design No. 35 may serve to give the perspective idea more justly.

These sketches show merely the geometrical design. It is a good plan so to arrange an arcade that it may be inclosed during the winter with glazed frames. This will make the house very much warmer during the cold months, and will help materially to protect the work from decay.

The *veranda* is perhaps the most specially American feature in a country house, and nothing can compensate for its absence. It may be constructed of lattice in the common way, or with a little more elaboration, as shown at A. A handsome veranda may always be made by using posts seven or eight inches in diameter, and fitting between them brack-

ets, or arches; and specimens of this style are shown at B and C. Another example, somewhat more elab-

orate, is shown at D. The veranda may, if preferred, be of more solid construction, an instance of which is sketched at E. An agreeable effect can be produced by turning the columns of some suitable pattern, and perhaps enriching them with a little carving. A very simple study for a veranda of this sort, lately designed for execution, is drawn at F. Wooden balustrades are of several sorts, as may be seen on the drawings of verandas just described. They may be constructed with turned

balusters, or composed of a flat surface, pierced with a pattern. A third method that may be adopted is somewhat of a combination of the other two. Its effect is produced by using flat balusters, as shown on parapet to C. Another sort is shown on the parapet to porch G, described already a few pages back, and endless opportunities for variety in design occur in treating this part of a country house.

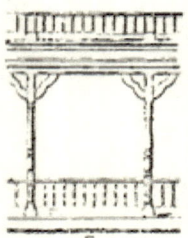

Windows are either single, double,

or single with side-lights. They may be fitted with outside shutter-blinds, which is the plan commonly adopted, as the cheapest and least liable to get out of order; or with inside Venetian blinds, or inside shutters, partly solid and partly filled in with slats, which is a very good way, or with altogether solid shutters. Inside shutters may be arranged either to slide or to shut up in boxes, as may be thought most expedient. French casements are troublesome if they have to open outward, and inconvenient, as far as curtains are concerned, if they open inward; but they have a pleasant effect in warm weather. In a room eleven feet high and over, almost the same advantage may be gained by sliding the lower sash up to the meeting rail, and inclosing it with a rough box and follower above, for the available height for exit will be five feet ten inches. The most perfect, and, of course, the most expensive arrangement, is to prepare a case in the wall sufficiently large to contain the sash, the Venetian blind, and the solid shutter, and then to slide each of these into this recess. All such work requires, however, to be very nicely fitted, and the blinds must be so planned that the slats will never be in the way. But these are matters of construction that it is scarcely worth while to go into very minutely here.

A few sketches of tin *ventilators* for bedrooms are drawn at M, N, O, and P. These are one tenth the

  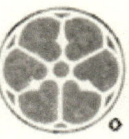 

price of registers, answer all the purpose (for a ventilator should never be closed), and look as well.

H

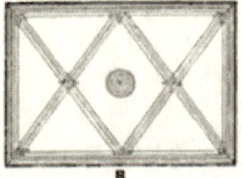 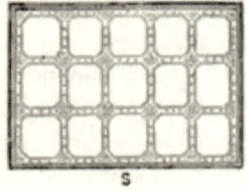

R   S

Some studies for *ornamental plaster ceilings*, that may be carried into execution by clever mechanics without much trouble or expense, are added at R, S, and T, and with these I conclude the illustrations and explanatory remarks that belong to this section of the work.

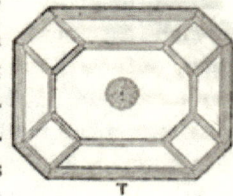

T

A point that requires much attention in the study of details is, to *make the ornament secondary to the construction, and not the construction secondary to the ornament.* This is the fatal rock on which so many a good conception for a house has been split. An inexperienced man, for example, may conclude to have an ornamental plaster-ceiling in his parlor, and in his desire to have it good of its sort, he may so load it down with decoration that it will be much more disagreeable to look at than the plain ceiling was before it was touched. And so it is throughout the whole subject of domestic architecture: *it is always as easy to spoil a house by overdoing it as by underdoing it.*

Having thus taken a passing glance at the most prominent points of detail that are likely to occur in the building of a country house, we may proceed to explain the various designs. But before doing so it is proper to remark, as so many different circumstances are likely to affect the arrangement of actual designs for country houses, that it would be useless for any architect to publish plans with the idea that they could be completely followed on any other site than the one for which they are specially designed. But this is not the object proposed to be gained. The principles they involve, and the individual peculiarities they possess can be combined, modified, and improved in many different ways. They are to be considered merely as hints or suggestions.

The vignette at page 25 illustrates a design for a village school-house, of which the plan is here submitted. It is not requisite that it should be of any particular dimensions, provided that the proportion shown between the different parts is duly observed. The outline of its plan is a simple parallelogram 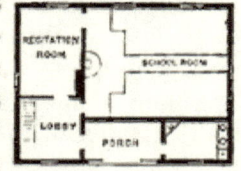 covered by one roof. The accommodation embraces a veranda porch, a lobby for hats and coats, a school-room, a recitation-room, a necessary, a loft overhead for storage of benches, etc., and a partially-excavated basement for a stove or furnace. A ventilator is proposed to be constructed on the ridge of the roof communicating with the school-room. The recitation-room would be ventilated by a spare flue alongside of the furnace smoke-flue. Such a building ought to be constructed of brick or rough stone, for it seems a pity to erect a school-house of wood, particularly in a country where whittling is almost an institution. Still, the design could be easily constructed of wood if it were thought advisable. A school-house of this sort could be erected for from $500 upward, according to size and finish required.

The vignette that closes this chapter shows a general view of a study made for the Associate Reformed Church proposed to be erected in Newburgh. This design is adapted for execution in brick and dressed brown stone, or the wall surface may be built with small, neatly-pointed, rough stones, and the window-dressings, angles, etc., may be of hard Jersey stone, fairly worked. The spire is proposed to be constructed of wood, above the octagonal portion of the tower. A more varied outline is thus obtainable; and as the main building is to have a roof and external cornice of wood, a similar principle of construction is thus carried out in the tower also. In the interior, which is designed without columns or galleries, a collar-beam roof is proposed to span the whole width of the building, and to show its construction in the arrangement of the panels of the plaster ceiling, each principal rafter being stiffened by a semicircular moulded arch springing on each side from a projecting stone corbel, prepared to receive it when carrying up the walls, which are intended to be strengthened at these points by buttresses, as shown on the sketch.

The want of taste that has been exhibited in the majority of designs for country churches in America has had a very prejudicial effect on the general appearance of villages and provincial towns; and many years must elapse, even under the most favorable circumstances, before a material change for the better can take place, for these structures are, in most cases, strongly built, and have cost much money. A wooden caricature of a Grecian temple has been the most popular form adopted, and this is repeated in a thousand meaningless, ugly ways all over the country. Here and there some slight appreciation of the beauty and appropriateness of the spire has been shown, but the examples are few and far between, and, when they do occur, are generally painted a glaring white from roof to base-board. It is the soaring, heavenward-pointing spire, bold in outline and quiet in color, that is the true architectural type of the spiritual Christian worship. This is the feature that should belong specially to the religious edifice, and thereby distinguish it at once from the secular buildings in its neighborhood. Its value and beauty, in connection with natural landscape, is, perhaps, unexplainable; but it is none the less real because we may be unable to trace in words the delicate process by which the mind perceives the propriety of its expression.

Since the first edition of this volume was prepared for publication, at Newburgh, on the Hudson, a church of considerable size, and with a lofty spire, has been built on one of the principal streets from the designs of my friend Mr. F. C. Withers; the whole exterior of the church, including the spire, is built of stone, and the general effect is shown in the vignette introduced at the end of the Table of Contents of the present edition. The body of the work is constructed of a grayish bluestone, the dressings and the spire are of a light olive freestone, and the roof is covered with two tints of slate; the color is thus diversified, although the contrasts are not so marked as to interfere with the general impression, which should always depend more on picturesqueness of line, and depth of light and shade, than on variety of local color in the materials used.

The question of an alteration in the appearance of country churches generally, is one in which we are all personally interested, exclusive of sect or denomination; for we are compelled to look at their outsides, although we are perfectly free to attend or stay away from the services that are held within them. And it is to be earnestly desired that the improving taste for good proportion and color that is perceptible in American villa and cottage architecture may soon reach this subject of country churches in the rural districts; for the building which we set apart for the service of religion, and in which we dedicate our best hours to the praise and worship of the good God who has created and preserved us, should surely be the purest, the noblest, and the *best* architectural work our minds can conceive and our hands execute.

DESIGN FOR A CHURCH.

DESIGN No. 1.—(V. & W.)

PERSPECTIVE VIEW.

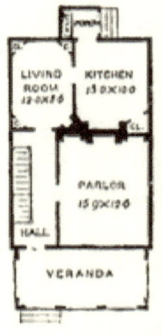

PLAN OF PRINCIPAL FLOOR.

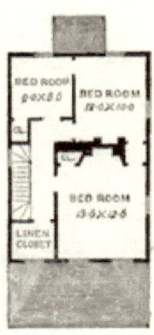

PLAN OF CHAMBERS.

DESIGN No. 1.
(V. & W.)

A SIMPLE SUBURBAN COTTAGE.

No. 1 illustrates a design for a simple rectangular cottage intended to be built on a twenty-five feet lot by Mr. Ryan, plumber, of Newburgh. This house is proposed to be constructed of wood filled in with brick, and to cost about $1500 or $1600, according to the amount of interior finish that may be decided on. It will be perceived by the plan that the house is approached by a veranda-porch, the principal floor being occupied by an airy parlor fitted with recessed bookcases in the walls on the side of the fire-place; a small living or dining room, and a kitchen communicating with a back porch, in which is a sink. The chimney-stack being designed in the centre of the building, all the heat from the fire-places will be saved to the house in winter. By this arrangement that objectionable feature of any plan for a house—a basement dining-room—is avoided, and the lowest floor is used for cellarage only, and not finished off. It would probably be convenient in the summer, however, to use a portion as a wash-room, and if so, a stove could easily be fitted up temporarily for the purpose. The chamber plan shows three bedrooms and a large linen-closet with a window in it. This, if preferred, could be used as a bath-room or a child's bedroom, and then it would probably be desirable to connect it by a door with the principal bedroom. The gable introduced at the side is for the specific purpose of getting a proper headway to the attic staircase.

It is not proposed, at present, to finish off the attic chambers; but the roof is of such a pitch that two light, airy bedrooms could at any time be obtained at a little additional cost, and without occupying the whole of the garret space. The house, having a twenty feet front, allows of two feet for projection of roof on one side of the lot, and of a three feet passage-way on the other. Another side-entrance at the head of the cellar stairs could be arranged, if thought worth while, for this three feet passage-way, but it is not in the present instance thought necessary. In fitting up the parlor of such a cottage as this, good taste would seem to indicate that the carpet should be of small pattern, and rather quiet in color, so as to give an air of repose to the whole room. The wood-work might be either stained and varnished, or painted in light, cheerful tints. The walls should be covered with a pretty, fanciful paper, harmonizing with the wood-work, and not large in pattern, or it would appear to decrease the size of the apartment. The mantle-piece may be of wood, of some simple, tasteful design, corresponding with the rest of the room, and yet look far better than a cold, costly white marble affair, that will run away with much money to no purpose. The centre-table should be a sensible, substantial piece of furniture, at which three or four people will be able to sit and read comfortably. A well-made chintz-covered lounge, although a much more economical, and a far more comfortable piece of furniture than a modern rose-wood sofa, will be found to have an equally agreeable effect in the room. Two or three tables of fanciful design and trifling expense, that can be moved wherever they may be wanted at a moment's notice, will give life and animation to such a parlor; and an easy-chair or two for tired visitors, besides the regular half dozen, will be found very acceptable. Some pretty, simple engravings on the wall in neat frames, and an oil-painting or two, can be obtained at a very moderate cost. Pretty casts for the mantle-piece, or to be placed on brackets here and there on the walls, may be obtained for a mere trifle. . I purchased one, for example, the other day, in New York, for twenty-five cents, full of grace, beauty, and artistic thought. A bird-cage, or a basin of gold-fish, or a hanging-basket for flowers, if there are any young girls in the family, will also help to give an air of vitality to the whole room, which should be the central point of attraction for all the inmates. It is possible, however, if we lay much stress on these minor accessories, that some Mr. Blank may say, "This will never do. We can't have our girls fussing around with flowers, and birds, and gold-fish. They have their duties to perform, and their studies to attend to." We will, therefore, stop here, merely venturing to remark, with all due deference, that although duties must, of course, be performed, yet innocent pleasures ought also to be encouraged, and that no study will insure so rich a reward to all concerned as the study of simple, quiet, domestic grace and elegance.

The vignette is intended to show a simple method of obtaining, in a new clearing, a comparatively comfortable and somewhat homelike family residence without much trouble or expense. This design does not illustrate a log-*cabin*, or single room, in which a family of men, women, and children eat, drink, sleep, wash, dress, and undress all together. It is a plan for a very simple house for a well-to-do settler and his family. The principal apartment, 16 × 20, is proposed to communicate at once with the open air through a door under a veranda-porch in the summer, and to be approached through a small wash-room at the side in the winter, the veranda being then used for storing a supply of wood under cover. In the wash-room is a flight of ladder steps leading to the loft. The family-room has two windows in it, and is connected with two small bedrooms and a store-room, each supplied with one window. The house is intended to be constructed in the ordinary manner with rough logs; but as much neatness as is compatible with proper economy is supposed to be exercised in the selection of the material and in the execution of the work. It is the common practice to cut down all the trees near the site of a log-cabin; but this custom is far more honored in the breach than in the observance, and a little judicious forethought will certainly preserve a few fine specimens round the family home for shade and enjoyment. This study has been made on a small, simple scale, so as to be more generally available, but the mode of construction proposed to be adopted admits of considerable artistic treatment in a rural way; and if a log-house were required on a more extensive plan altogether, the style might be raised in character accordingly, without sacrificing in any way its primitive expression. Log-houses are frequently occupied, for years together, by well-educated, active, energetic men, who are the pioneers of civilization in the thinly-inhabited districts in which they take up their abode; and much good would result if such men would set the example of devoting some thought to the *beauty*, as well as to the *utility* of the homes they erect for themselves and their families.

We are all aware that a decided distinction is commonly drawn between the two words I have just used, viz., beauty and utility; and yet every useful object should be beautiful, and every beautiful object must be useful, or its alleged beauty is hollow and unreal. The misconception arises from the shallow material tests with which we are apt to content ourselves. A glorious sunset, for example, has nothing whatever to do with eating, drinking, sleeping, or locomotion, and it may, therefore, be said by some to be an entirely useless affair; but the fact is, that, if we are wise enough to appreciate it, the beautiful sunset is of great use to our *higher*, although it is of no use to our *lower* capacities of conception. A horse may be made to lead a life of hard work, and may enjoy his warm coat, his victuals and drink, and his sound sleep, as well as a man; and a human being who devotes his life solely to obtaining these acknowledged necessaries of his existence, will never be any thing better than a rather superior specimen of a beast of burden. But no beast of burden has yet evinced any special partiality for sunsets, or has shown any natural craving for beauty for its own sake. Here, in fact, we arrive at one of the main characteristics that especially *distinguishes* the man from the brute. Up to this point, all the enjoyments his five senses are capable of he shares with the inferior animals. But the capacity to see and fully appreciate the beautiful, and to discover the charm and wonderful excellence of loveliness, is restricted to the human race. *Man alone*, of all living beings on the earth, has been made in the image of his Maker, and man alone can experience the delightful sensation of sympathy with the divine attributes of grace and beauty. How debasing, then, must be that leveling, rational, utilitarian spirit, that seeks to cut out from the desires and aims of men all that is useful to them as beings who are pronounced by their Creator only a little lower than the angels, while it insists on an almost sleepless attention to those material cravings that are useful to them, simply because they are active, energetic animals. All this eating, drinking, sleeping, and working for means of subsistence is not "life," although it is a needful preparatory step to it, and may be made an enjoyable part of it. It is to life what the foundation is to the building: it may be excellent in itself, and deserves, undoubtedly, the most careful attention, because a flaw in this part is fatal to all that may be done afterward. But what opinion should we form of that man who spent all his time and money in building foundations? The true way to live, says the prudent economist, is to pay as we go: and this rule is of thousand-fold application. If we wish to realize our existence, we must pay as we go, not only our debts to the body, but what we owe to our higher, purer, better, and more ideal nature, or we accumulate a heavy debt that drags us down in after life. Some people form a notion that they can devote five, ten, fifteen, or twenty years to the accumulation of means to purchase enjoyment, and can then sit down comfortably and enjoy it. But the order of nature is entirely averse to any such investment of time. She permits nothing of the sort. Pay as you go, says Nature; clear up accounts every day with your good genius, and cheerfully subscribe something to your ideal life, to your taste for the beautiful, to your domestic happiness, or, when the ten years are passed, you may find yourself with a large account at the bank, but with very little capacity to enjoy any thing that your money can purchase. And it is not for ourselves alone, but for the sake of our children, that we should love to build our homes, whether they be villas, cottages, or log-houses, beautifully and well. Men and women can go abroad and take their pleasure elsewhere. They can gratify their desire for variety and excitement in a hundred different ways; but the young people are mostly at home; it is their storehouse for amusement, their opportunity for relaxation, their main resource; and thus they are exposed to its influence for good or evil unceasingly; their pliable, susceptible minds take in its whole expression with the fullest possible force, and with unerring accuracy. They soon learn thoroughly to enjoy every possibility of enjoyment it possesses, and their unspoiled instincts for the good and true are perpetually seeking in it for a gratification of their nascent perceptions of the beautiful. It is only by degrees that the young, hungry soul, born and bred in a hard, unlovely home, accepts the coarse fate to which, not the *poverty*, but the *indifference* of its parents condemns it. It is many, many years before the irrepressible longing becomes utterly hopeless; perhaps it is never crushed out entirely; but it is so stupefied by slow degrees into despairing stagnation, if a perpetually recurring blank surround it, that it often seems to die, and to make no sign; the meagre, joyless, torpid home atmosphere in which it is forced to vegetate absolutely starves it out; and thus the good intention that the all-wise Creator had in view, when instilling a desire for the *beautiful* into the life of the infant, is painfully frustrated. It is frequently from this cause, and from this alone, that an impulsive, high-spirited, light-hearted boy will dwindle by degrees into a sharp, shrewd, narrow-minded, and selfish youth; from thence again into a prudent, hard, and horny manhood, and at last into a covetous, unloving, and unloved old age. The single explanation is all sufficient: he never had a pleasant home.

DESIGN FOR A LOG-HOUSE.

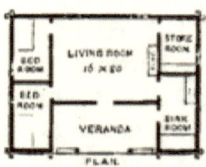

DESIGN No. 2.

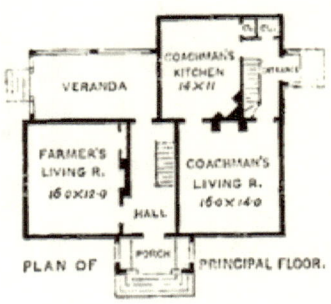

PLAN OF PRINCIPAL FLOOR.

DESIGN No. 2.

A SMALL RURAL DOUBLE COTTAGE.

No. 2 is a design for a cottage for two small families. Under the farmer's living-room is a basement-kitchen, with the windows considerably out of ground, and under the coachman's kitchen is a cellar: the entrances are, as will be perceived, quite distinct. Up stairs the farmer has three bedrooms; the coachman but one, according to instructions. Such a cottage would cost about $1800, neatly finished.

Such a plan would not be unsuitable for a lodge, in which the families of a gardener and gate-keeper could live, or it might be fitted up a little more completely, and offer convenient accommodation to two friends who felt inclined to build it on some agreeable rural lot for a few months' quiet residence in the summer.

It seems strange that this idea should not be more frequently acted on than is the case at present. Far away from the fashionable watering-places, but easily accessible from the cities—in the heart of Vermont, for instance—may be found bold, beautiful scenery, pure air, and a pleasant neighborhood. Land is cheap, timber cheap, living cheap, and all of the best. These are the spots that should attract the attention of heads of families who wish to give their young people the benefit of country life in the summer. A long trip of three months may, in this way, be taken at a less cost than will be incurred for a brief, glittering three weeks at Saratoga or Newport, and with real, instead of nominal, advantage to the health of the juniors who join in it.

This cottage is proposed to be constructed of wood, filled in with brick, and covered with clap-boards. The upper view shows a rear addition to the coachman's part of the house, which is not indicated on the plan, and which might be included in the design, if thought worth while, although the house is complete without it. The verge-boards and porch are proposed to be slightly ornamented, and a hooded door, adapted to the arrangement shown on the ground plan, is illustrated in the vignette below.

DESIGN FOR A HOODED DOOR.

DESIGN No. 3.

PERSPECTIVE VIEW.

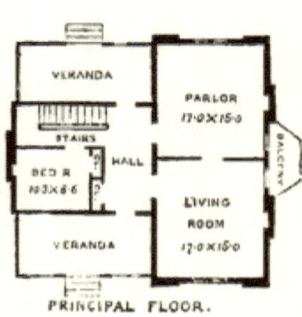

PRINCIPAL FLOOR.

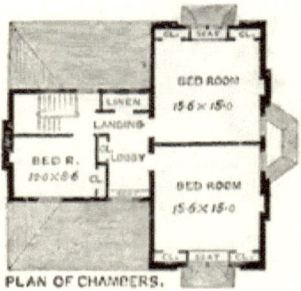

PLAN OF CHAMBERS.

DESIGN No. 3.

A SUBURBAN COTTAGE.

This design for a suburban cottage on a small scale was prepared for a situation on a street lot, in which the house would have been generally seen among trees, and in connection with the other houses adjoining. The proportions were, therefore, made somewhat higher than would have been thought desirable if the site had been larger and more open. This point of *relative* proportion is worth much consideration in suburban houses, for it may easily happen that a particular design shall have a decidedly dwarfish appearance if built in one situation, a high, stilted effect in another, and be quite satisfactory in a third—the result on the eye being dependent as much on the objects immediately surrounding the house as on the house itself. The force of this remark may be tested

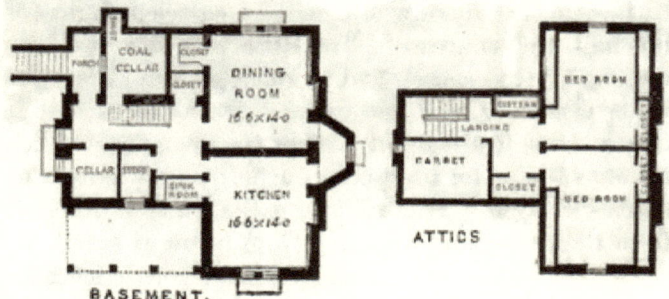

convincingly in any large city by an examination of the architecture of its shops and warehouses. It frequently happens that a block of stores, which for years has presented a tall, massive, dignified appearance, suddenly shrinks into stunted, second-rate insignificance, simply because other buildings two or three stories higher are erected on each side of it.

In accordance with instructions the dining-room was planned in the basement. It connects with the kitchen through a pantry that is arranged under a balcony on the principal floor, and is supplied with a large china-closet. A sink-room is attached to the kitchen, and the remainder of the basement is occupied with cellars and store-room.

The principal floor shows a veranda-porch communicating with a hall, in which is a cloak-closet and door to back veranda. The stairs are in a side hall, in which is an entrance to a bedroom of moderate size; and a parlor, connected with a general living-room, and opening on to a balcony, completes the accommodation on this floor. The chamber plan shows three bedrooms fitted up with closets, etc., and a linen-press in the hall. If preferred, the other end of the open hall could be inclosed and fitted up as a bath-room, for there would still be sufficient light for the hall and staircase. The attic provides two bedrooms and store-closet, and an open garret. The rain-water cistern, it will be observed, is so placed that it may be sunk in the upper part of the linen-press, which is not available for its special purpose more than six or seven feet high. By this means the water is received from the eaves, and pumping from below is saved.

In this plan, as it interfered with convenience to arrange the chimneys in the body of the house, they are placed in the outside walls, and are intended to improve the appearance of the design, although perfectly simple in execution. The chimney is a most expressive feature, and deserves to be brought prominently into notice in domestic architecture. As a general rule, it is desirable in this climate to build the chimneys in the body of the house, and not in the outside walls. But exceptions often occur in large houses, and sometimes in small ones, where the gain in so doing is greater than the loss, and in such cases the opportunity to give a definite character to the chimney-stacks should be taken advantage of. This design has not been executed. It was estimated by Newburgh mechanics at over $4000, in 1852; but as it is not nearly so large a house as Design No. 7, which was built, to the satisfaction of the proprietor, for $4200, complete, about twenty miles from the site on which this design was proposed to be erected, it is presumed that, under favorable circumstances, the estimate would be within that amount. It was proposed to be built of brick, furred off on the inside, and painted a soft cream color externally, the verandas and trimmings being finished a rich brown.

In arranging the tints for the exterior of a country house it is better to make them a little warmer than will be entirely satisfactory at first, because the pigments must certainly fade more or less in a few months, and the permanent effect is what should be most thought of at starting.

The vignette shows a design for partially inclosing a veranda which was made for J. J. Monell, Esq. The country house to which it belongs is planned on a side-hill overlooking a picturesque glen; and the kitchen in the basement being, in consequence, entirely out of ground in the rear facing the garden, these offices were somewhat too freely exposed to view from the ornamental grounds in the vicinity of the house. The simple plan here sketched was therefore made and carried out; and as the vines, already grown over the old supporting-posts, were trained to the new trellis-work, and showed to even better advantage than before, no harm was done in this respect, while a greater degree of privacy was certainly arrived at with but little sacrifice of light. The sketch is introduced as similar cases often occur, and a little ingenuity can readily arrange new varieties of pattern to embody the same general idea.

DESIGN FOR PARTIALLY INCLOSED VERANDA.

DESIGN No. 4.

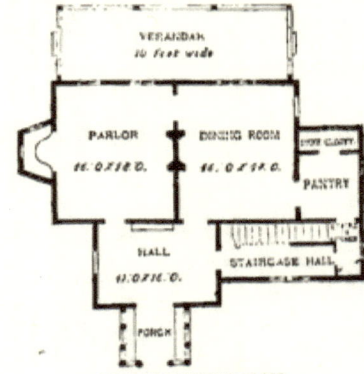

PLAN OF PRINCIPAL FLOOR.

DESIGN No. 4.

RURAL COTTAGE.

This design, erected by Dr. de la Montagnie, of Fishkill Landing, is situated amidst quiet, agreeable home scenery, and commands several beautiful views, both of the Hudson and of the noble hills that rise up at this point from its eastern shore. The approach road near it is picturesquely wooded on both sides, so as to seem more like a wide lane than a common highroad; and all the circumstances suggested an unpretending, but really rural house.

As the accommodation required was not extensive, there was no necessity for attic bedrooms, and the cottage is, therefore, planned a story and a half high, as it is called, the roof coming down a foot or two below the ceiling line, not, however, so as to interfere with the occupation of the rooms in any way. Although, by this arrangement, the ceilings of the bedrooms are less elegant than they would be if finished off square, as usual, there is one decided advantage attending it,

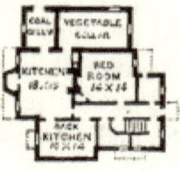 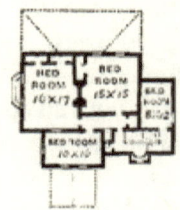 

which was fairly illustrated in the case under consideration. The ground in the immediate vicinity of the building site was, as is often the case, somewhat bare of trees, and the proprietor, with great care and pains, moved a number of healthy specimens, of larger size than usual, from the neighboring woods. Fortunately these have thriven well for the most part, and the consequence is, that although they are small, and have had only a year or two's growth in their present situation, they have quite an important effect in connection with the house, *because it is kept low*, and with overhanging eaves that still further take away the effect of height.

This design was built in a hollow, but the earth taken out for foundations was so arranged that the house, as now finished, stands on a gentle eminence, and the natural impression of a stranger to the facts would be that Nature kindly provided a little bluff for the specific purpose of building the doctor's cottage on. The general effect was materially assisted by sodding the surface in the vicinity of the house, instead of trusting to grass seed. This process is, of course, the more expensive of the two, but if well done, it yields, what is really of importance in a new house, an immediate reward.

The entrance is through a wooden porch, that serves for a veranda on that side. The hall is of liberal size, and is almost as useful as another room, having a window in it commanding a pretty view.

The parlor and dining-room communicate with the veranda. A roomy pantry and lock-up closet are provided in connection with the dining-room, and a lobby is shut off at the head of the basement stairs: in this is a wash-stand, etc. Up stairs are four bedrooms. There is a large garret of good height overhead. The kitchen accommodation, etc., is in the basement.

In the arrangement of the chamber plan it will be

seen that a projecting dormer-window is introduced in the upper hall at the head of the stairs. This supplies a pleasant recess for a chair and table at a window that looks out on a cheerful view; and as it stands out from the main hall some little distance, being supported on heavy brackets, as may be seen on the upper illustration, it casts a deep shadow, even when the sun is high, and gives some additional individuality to the design.

The house is constructed of wood, filled in with brick, and the carpenter's and mason's contracts were taken at $2900; the architect's commission for drawings and details, without superintendence, being 3½ per cent. on that amount, viz., $101 50.

One advantage that is offered by wooden construction is, that picturesque breaks in the plan may be made for less money than they will cost in brickwork, because it requires considerable time and care to make a brick corner plumb and true, while a wooden angle can be easily worked.

The exterior is painted in quiet, neutral tints, the main body of the work being of a rather warm gray, while the corner-boards, verge-boards, window-dressings, veranda, and porch are also of a grayish tint, but considerably darker than the other, and with some brown added to it for the sake of contrast. The stiles of the Venetian blinds are rather lighter than the window-dressings, while the slats and the panels of the verge-boards are of a cool dark brown. The chimney is painted in two tints, to correspond; and as the house is covered with shingles, which soon become soft and pleasant to the eye, the whole effect is free from either startling contrast or wearisome monotony.

In painting a country house the aim should be to

give it a cool effect in summer, and a warm effect in winter; and this is not so difficult as might at first be supposed, because all combinations of colors are mutually dependent on each other, and the marked contrast in the appearance of the surface of the soil, the trees, and the sky, at different periods of the year, gives an opportunity, when choosing the tints for a house, to select a happy medium that shall be suited to more seasons than one.

The vignette shows a sketch for a rustic outbuilding proposed to be constructed with a rough frame covered with bark, and with a shingle roof.

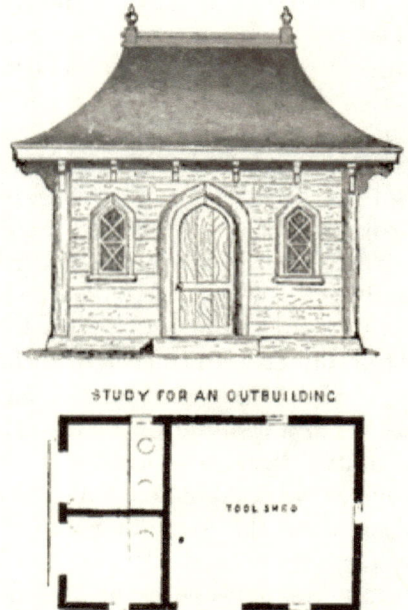

STUDY FOR AN OUTBUILDING

DESIGN No. 5.—(V. & W.)

PERSPECTIVE VIEW.

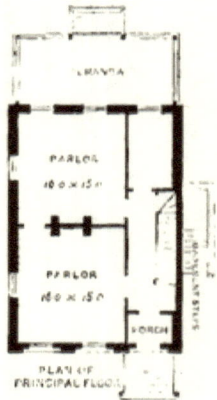

PLAN OF PRINCIPAL FLOOR.

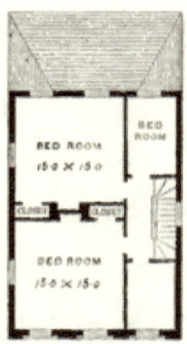

PLAN OF CHAMBERS.

DESIGN No. 5.
(V. & W.)
SUBURBAN HOUSE.

This is a design for a suburban residence for Rev. E. J. O'Reilly, of Newburgh. It seeks to supply the accommodation of an ordinary three-story brick house in a form that shall have a less high-shouldered and stilted appearance than usually distinguishes buildings of this class.

In the plan a recessed lobby is arranged, with outer door for shelter, instead of a projecting porch. The principal floor shows two parlors and a small study; or, if another arrangement is preferred, a parlor, a library, and a spare bedroom, the back rooms opening on to a veranda. In the basement is the dining-room, kitchen, coal-cellar, and pantries, with a side entrance communicating with the same. This arrangement is to be preferred to the more common plan of putting the kitchen door under the area steps, as it gives more privacy, and uses up less space in the basement for passage-ways. The bedroom plan supplies two large and two small bedrooms, one of which may be fitted up as a bath-room. The arrangement of the attics is somewhat similar to the chamber plan, two roomy bedrooms being supplied over the principal apartments, and garret and store-rooms over the remainder. These latter are lighted by small windows under the eaves, the other two by the windows in the gables. The roof projects two feet six inches all round, and is finished with a simple eaves moulding and brackets. The pitch of the roof is so high that there is a considerable space between the ceiling of the attic rooms and the flat overhead, and thus these rooms are not open to the objections that often apply to attic rooms in the hot summer weather. The front is paneled in brick, the panels being painted of a tint some shades darker than the rest of the house, to avoid monotony, and to give value to the brick projections. This house was built in a substantial manner, the carpenter's and mason's contracts being taken at $3000, including two coats of paint, but exclusive of plumbing and mantles.

The vignette shows plans and a view of a double cottage proposed to be erected on a fifty feet lot. No advantage whatever results from building two small detached cottages, with the same amount of accommodation in each, on two twenty-five feet lots; and by building them back to back one wall is saved, and both houses are rendered much drier and warmer.

The roof is simpler, and offers better attic rooms. The passage-ways at the side of each house are wider, and the whole effect is more dignified and agreeable. The plans are so drawn on the illustration that the entire accommodation is set forth, the principal and chamber floor being of one house, and the attic and basement of the other. This will require a little care in examination, to be thoroughly intelligible. Each house corresponds exactly with the other, except that the plans are reversed.

The accommodation provided may be thus described: A recessed porch, covered with a hood, leads to a hall, in which a staircase to the upper rooms is planned in the usual way. The parlor, which faces the front garden, is lighted by a bay-window, and communicates with the dining-room, which has windows to the floor opening on to a veranda. In this room a recess is prepared for a deep side-board, and a door at the farther end connects with a pantry. This is shut off from the main hall by a lobby at the head of the basement stairs, and is fitted up with closets, in connection with which a lift may be contrived, if required. Some few steps down this inclosed staircase is a landing, from which a door opens from the passage-way at the side of the house, thus securing a separate kitchen-entrance, and a convenient side access to the premises, without any sacrifice of space in the passage-way, and, at the same time, avoiding a servants' entrance down area steps, which is an advantage, as an area is likely to get filled with snow in the winter.

In the basement will be found a kitchen, with sink-room adjoining, the kitchen being provided with a large pantry, and a door to the furnace-room and coal-cellar. There is also a store-room and provision cellar.

In the chamber plan are two large bedrooms and a smaller one. The principal chamber communicates with a bath-room, which has another door from the hall; and near the bath-room is a water-closet, with a separate external window and a linen-press. In the attic three large bedrooms are planned, and a space for lumber is marked on the plan; but this might be used as a clothes-room, if preferred, and the space overhead between the attic ceiling and the flat could have a step-ladder to it, and be used as a lumber garret.

These houses might be built of eight-inch brick, furred off above the basement, the party wall being sixteen inches thick, so as to contain all the ventilating flues that may be required.

This block of two might be built of brick for $5000, with a simple internal finish; and it is much to be regretted that some attention is not bestowed on this class of buildings, as it is a more economical, and far preferable arrangement to erecting small, detached buildings within a few feet of each other, as is generally done on village streets. In this design the rooms are supposed to be of good size, and the whole arrangement is adapted to the requirements of a man of business in comfortable circumstances, who requires to be not farther from his office or store than the immediate outskirts of the country town in which he resides, and where, consequently, extra land will be both more valuable and of less use than it would be farther away from his neighbors. The same idea might, if preferred, be developed on a smaller and cheaper scale.

DESIGN FOR A DOUBLE COTTAGE.

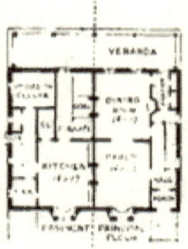
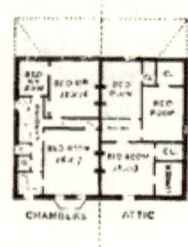

DESIGN No. 6.

PERSPECTIVE VIEW.

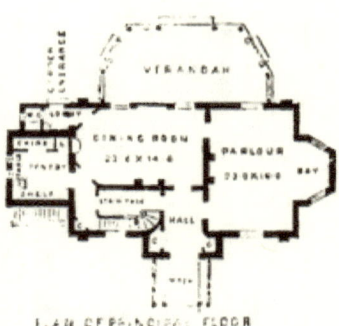

PLAN OF PRINCIPAL FLOOR.

DESIGN No. 6.

MODEL COTTAGE.

This design has been prepared to illustrate the views set forth in the opening chapter with regard to the real needs of a small family home in the country, and may be supposed to be arranged for the use of a man of simple habits, with some refinement of taste. Such a man would be a lover of hospitality, fond of fresh air, free space, and an easy life, and willing to take advantage of all reasonable modern improvements in the art of living comfortably; but, from choice as much as necessity, being economical in his requirements, he would have as strong an aversion for empty display as for scanty baldness in arrangement or detail.

The plan shows an entrance-porch and small hall, in which are hat and cloak closets. All the family accommodation provided on this floor is a parlor communicating with a dining-room, and both opening on to a wide veranda. The parlor has a large bay projection, with seat round it, and the dining-room is fit-

ted up with a book-case, B, a side-board recess, R, and connects with a roomy pantry, in which is a lift, L, a sink, S, a china-closet, a row of shelves, and a hanging-table. The dining-room also communicates with a garden entrance-lobby, fitted up with a washstand, and connecting with a water-closet. The staircase-hall is shut off from the main hall, and the basement staircase, opening on to the pantry, is partitioned off from the principal staircase; thus all necessary privacy is insured. The basement plan explains itself, cellarage being obtained by excavating under the veranda. The chamber plan supplies a family bedroom, with dressing-room, large closets, and bath-room attached: it also connects with a small bedroom that has an entrance from the hall. A water-closet is provided close to the bath-room; and one other guest's bedroom, as will be seen on the plan, and a linen-room under the roof of pantry building completes the accommodation on this floor. In the attic are two good bedrooms, a store-closet, and a large garret. Another bedroom might be finished off, if preferred.

The wooden outside porch is proposed to be finished with an open timber roof, the rafters being planed smooth and chamfered on the edge, and the boarding being matched and beaded. Provision is also made here for fitting sashes in the panels, and for hanging an outer door when required. Such a porch should be paved in preference to being boarded, and as permanent seats are planned on each side, and the projection from the house is ten feet, an arrangement of this sort will be found by the inmates a very tolerable substitute for a veranda when the sun is shining on the opposite side of the house.

The large bay projection in the parlor is proposed to be constructed of the same materials as the walls of the house, with three sash-windows fitted into it. The ceiling is intended to be of the same height as the room, and to have a balcony over it accessible from the chamber above. By finishing the bay projection without an interior arch, the apartment will be much increased in apparent dimensions, as the eye is naturally apt to judge of the size of a room by the boundary lines of the cornice. There are several methods of treating such an arrangement of a bay projection in a satisfactory manner internally.

The other living-room, which should be library and dining-room in one, might appropriately be finished with Georgia pine, unpainted, and a moulded skirting, or wainscoting, about two feet three inches high round the sides, would connect the various openings together in an agreeable and not very expensive manner.

The exterior of this house is simply designed. There is no attempt to make it "all corners and all gables," or to evade the fact that it is a straightforward, rectangular house; but, at the same time, an effort has been made to give it a somewhat picturesque character in the arrangement of the roof and dormer-windows. The slight curve at the eaves (of which the method of construction has been explained in the introductory chapter) helps materially to reduce the hardness and angularity of line that would otherwise be somewhat prominent in such a simple arrangement of roof. It will also be observed that, as the principal gable is wider than the entrance hall, the parlor window, which is naturally planned in the centre of the room, would appear one-sided on the exterior elevation, if some precaution were not taken to avoid this disagreeable effect. A slightly-projecting pier, arranged to receive the rain-water pipe, is therefore introduced at the angle, and the balance is thus restored, as the windows appear now to be planned in the centre of the panel thus formed, and a corresponding arrangement is, of course, adopted on the staircase side of the hall, so that the necessary uniformity may be preserved.

Such a house as this could be fairly built for $3500 in ordinary building times, and is equally fitted for a brick, wood, or stone construction. The same general plan may be amplified so as to include a kitchen wing with back staircase, and may be adapted to a flatter pitch of roof. An example of such a modification, which was prepared for a site in the neighborhood of Shrewsbury, L. I., is illustrated below.

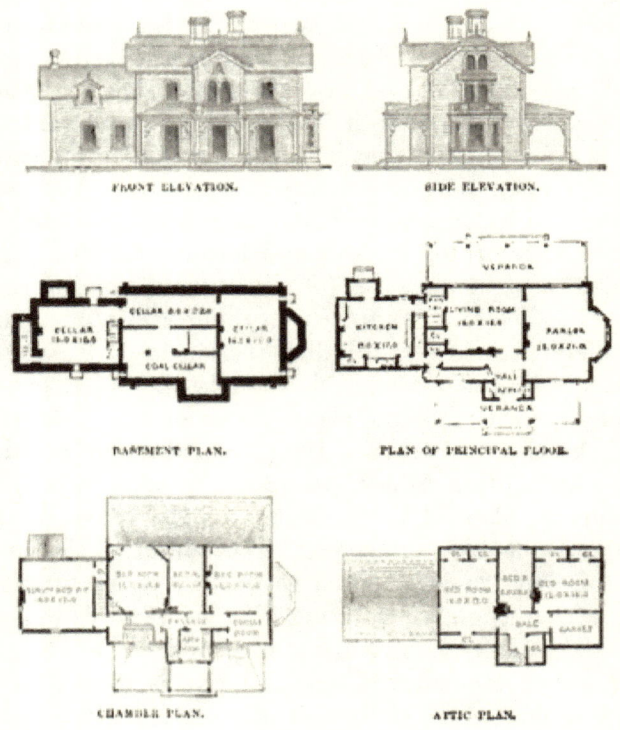

FRONT ELEVATION.   SIDE ELEVATION.

BASEMENT PLAN.   PLAN OF PRINCIPAL FLOOR.

CHAMBER PLAN.   ATTIC PLAN.

The vignette shows a study for a farm-house intended to be built of brick or stone. The aim here has been to design a building that shall be domestic and simple, and yet not unsuited in artistic effect to take its place in a rural landscape. One disagreeable peculiarity that is often noticeable in American farm-houses is, that they are too full of windows, and have, in consequence, an undignified, mean effect. Breadth of surface has been sought for in this study, and no more windows have been introduced than entire convenience requires, the plan being arranged accordingly. Thus, though there are four openings under the front veranda, only two are inserted in the next story; for if two more windows were added on the chamber floor, the whole effect of repose would be destroyed, without any advantage being gained in interior comfort. The accommodation in the main part of the house consists of a hall with a staircase in it, a parlor communicating with a general living-room, and a bedroom connecting with this apartment and the kitchen wing. It is not thought necessary to provide a separate passage to the kitchen from the front door, and it is calculated that the family-room will be used as a dining-room. It is the custom with some farmers to make a constant practice of taking all meals in the kitchen; but this habit marks a low state of civilization. The occupation of farming is the natural employment of a human being, and it ought to be made a refined and noble pursuit, not a mere way of earning a rude subsistence. It is among the sons and daughters of the farmers that the pith and marrow of a country are to be found, and every grace that belongs to rural life should find its highest examples in the home and family of the intelligent American farmer. The wing building in the design under consideration contains a side entrance, with veranda-porch, several pantries, a roomy kitchen, a wash-room, and a wood-house. The chamber plan in the main building contains three large and two small bedrooms, a linen-press, and a staircase to an extensive open garret. The kitchen wing contains four secondary bedrooms, approached by a flight of stairs in the wood-house, and accessible also from the main building through one of the bedrooms, as it is not thought worth while in a farm house to sacrifice the space that would be required for a separate communicating passage between the upper hall in the main house and the wing rooms.

DESIGN FOR A FARM-HOUSE.

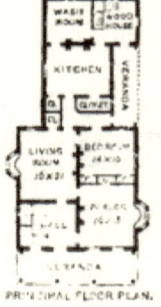

PRINCIPAL FLOOR PLAN.

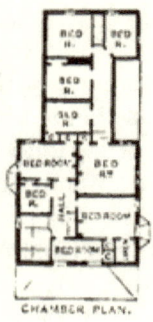

CHAMBER PLAN.

DESIGN No. 7.

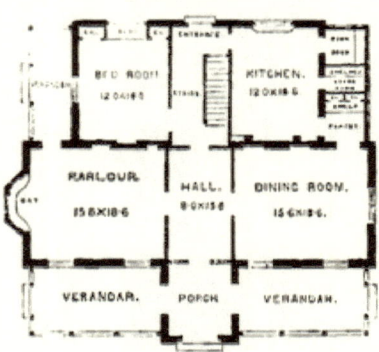

DESIGN No. 7.

COTTAGE RESIDENCE.

This design illustrates a cottage residence built for Mr. Wright, of Goshen. It was required that the kitchen and its accessories should be on the same floor as the living-rooms, but the house was not proposed to be of such a size as would warrant the erection of a separate wing for this purpose. The whole building is, therefore, under one roof, and the kitchen is so placed that its contiguity to the principal rooms does not interfere with the privacy that properly belongs to the apartments in constant use by the family.

A porch of brick, communicating by arched openings, with verandas on each side of it, forms the principal entrance, and opens on to a hall 8 × 15 feet six inches. This porch is so arranged that the arched openings at the sides can be closed with glazed frames in winter, and the central opening fitted with a frame and door, thus making a double hall, that is a great

 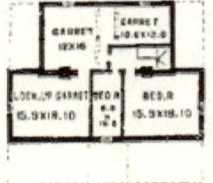 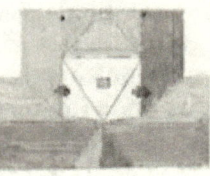

advantage in severe weather, as it prevents the ingress of a draught of cold air whenever the front door is opened, and offers a protection from storm to visitors while waiting for the servant to attend to the bell.

The parlor and dining-room open from the hall by

L

doors opposite each other. The parlor has a bay-window and a door opening on to a small private veranda that is not overlooked by any one approaching the house. The dining-room is 15 0 × 18 6. It is undesirable that any dining-room, and more particularly one that is used as an ordinary living-room, should be directly connected with the kitchen, for various evident reasons; and it is equally inconvenient to have the halls and passages that belong to the other apartments occupied several times a day by the servant whose business it is to prepare the table for meals and to clear away afterward. In the present instance the difficulty has been overcome by constructing a one-story inclosure corresponding in design with the private veranda on the other side of the house. This is of wood, and, at small cost, adds much to the convenience, and something to the appearance of the house. Its effect is shown on the lower perspective view. A pantry between the kitchen and dining-room is thus obtained, communicating with both; and by this means convenience of access, without loss of privacy, is secured. A store-room for the kitchen is also supplied in this little outbuilding, and also a sink-room; this latter, having two small windows on opposite sides, is well ventilated, and renders the kitchen a much more pleasant apartment for servants to live in, as it relieves it of the most disagreeable part of the work. The kitchen is 12 0 × 15 6, and has a door close to the back entrance and to the cellar stairs. If preferred, this back entrance might be shut off entirely from the principal staircase by a door across the hall on the same line as the cellar door, thus disconnecting the kitchen still more completely; but it is not shown so on the plan.

The staircase hall is entered from the main hall, and communicates with a bedroom, 12 × 14, on the principal floor. This room could, of course, be used as a library or study, if preferred; but it adds much to the convenience of a country house to have one bedroom that, in cases of sickness, can be approached without the labor of going up and down stairs. It is worthy of consideration in a cottage residence that, at seasons of festivity, the rooms may be required to hold an unusual number of persons; and though it is not desirable to sacrifice any family comfort in the plan on this ground, it is as well to bear it in mind. In this design, in the event of an entertainment, the rear entrance should be used by visitors; the bedroom would thus be close at hand, and could be used as a cloak-room, while the parlor and dining-room would make one *suite* with the front hall and porch, the doors of communication being thrown open. Additional room could be easily gained for a summer entertainment by temporarily inclosing the front veranda outside with calico, or any similar material, lightly affixed; and if the veranda posts are covered with creeping plants, as they should be, the effect of such a leafy gallery as is thus obtained when lighted up at night is very cool and elegant.

The chamber plan contains four bedrooms and a small study or sewing-room, 8 × 11, opening by glazed doors on to a balcony over the porch. A flat of this sort offers a good opportunity to the ladies of the house to cultivate flowers in pots with little trouble, and when thus used, it is a useful and agreeable accessory. In this country, either from climate or some other cause, ladies generally find it either too hot or too cold for gardening; and as it devolves on the feminine portion of the household to attend to these minor matters where a regular gardener is not engaged, it is worth while to take advantage of any opportunity like this to keep the flower vases filled with finer and better-tended specimens than will be likely to be found in the flower-beds on the lawn.

The attic plan contains two servants' bedrooms and a roomy garret.

The plan of the roof of this house will illustrate the general arrangement that experience seems to show is the most desirable for the roofs of country houses exposed to a climate so peculiar as that of these Northern States: it is suited, indeed, for any climate, but is especially called for here.

The main requirements in a roof are, that it shall keep off the wet; that it shall offer no opportunities for snow to lie where it can thaw, and freeze, and be a nuisance; and that it shall supply cool and commodious attic rooms, if needed. By projecting the roof two feet six inches or three feet, and cutting the gutter into the rafters, the walls will be well protected from the wet, and by invariably avoiding solid parapets and horizontal gutters between gables, the snow will be no inconvenience. A flat connecting the various ridges supplies free room in the attic, and is in every respect a complete mode of construction, while any plan that requires gutters between the gables is sure to give trouble at some time or another. The snow that falls on a flat of this sort is generally blown off at once, as it is so much exposed to the wind. If, however, it should happen to lie, and thaw and freeze for a month together, it can not do any harm, as it is unrestricted on every side, and drips away down the roof as it melts.

Very much depends on the appearance of the roof in a country house, for it is the first and last feature that impresses the eye, and it should be made an important part of every design. It is for this reason that so many plans of roofs are inserted in this work, and I hope that the casual reader may be induced to give some attention to their arrangement. For this design the plans, specifications, and working drawings were furnished without superintendence; and as the house has been erected several years, I lately wrote to the proprietor, asking him as to its ultimate cost, and if a personal residence in the house had suggested to him any improvement on the plan. He replies, "As near as I could get at the cost of the house, for which you drew plans, it was $4200. With reference to alterations in the plan, I know of none that I would be willing to advise, although others of more knowledge of construction might."

The vignette illustrates an artist's studio, designed for Mr. Jervis M'Entee, landscape painter, of Rondout, and built by him of wood, fitted in a substantial, simple way, for $400. It is finely placed on an elevated site, and commands an extended view of the Kaatskills and the Hudson. On the plan it is one simple room: the ceiling line follows the line of the roof and collar beams, so as to give height and a more airy effect to the interior. All the rafters are left visible, the plastering being fitted between them. Some time after the design had been built and occupied, Mr. M'Entee added a portion of a simple cottage residence in corresponding style, the studio still remaining in use for its original purpose. The complete effect aimed at is shown on the lower view and plan.

As the parlor, the veranda, and the porch are not yet built, a somewhat disproportioned result is obtained *for the present,* as the wing looks larger than the house; but its accommodation and cost, as it stands, is all sufficient for the immediate needs and circumstances of its proprietor, who has judiciously preferred to run the gauntlet of his neighbors' criticism for a time, and to plan his house as he *will* want it, carrying it out by degrees as opportunity offers, rather than to adopt a snug arrangement complete in itself, which, although suitable enough for to-day, would, in all probability, be in a few years inconveniently small for his needs; and if he wanted to sell at any time, could hardly fail to prove an undesirable investment on a site constantly improving in value, and that might, in all probability, be disposed of to advantage at any time, with a roomy house on it, or a building that could easily be made a good family residence, without pulling down the existing building or injuring its general appearance. The house as it at present stands, with hall, dining-room, pantry, small bedroom, and studio, on principal floor, three bedrooms and a little bath-room above, and basement kitchen, with cellars below, has cost $2000; and Mr. M'Entee calculates that another $1000 would render it complete, as shown, giving a second kitchen below, a best parlor and a best bedroom over, in addition to the accommodation already provided.

The scenery in the neighborhood of this cottage is of the most striking and varied description, and the eye looks over a range of country extending from the North and South Beacons at Fishkill, on one side, to the lofty Round Top and Woodstock Peak on the other. A white, isolated speck, visible in the gray distance, marks the situation of the Kaatskill Mountain House, and calls to mind the many beautiful spots in its vicinity, one of which, "The Falls," I have thought it worth while to speak of more particularly here, because its pictorial effect is much injured by the unarchitectural treatment it has received at the hands of those who have attempted to make the access to it more satisfactory to tourists. This beautiful fall of water is so picturesquely grouped in connection with the trees and the rocky precipitous sides of the mountain, that its upper edge, or lip, seen through the spray from below, would appear to be at a great height from the eye, if it were not for a clumsy boarded structure that has been erected just on the brink of the descent to afford visitors a view clear down into the valley. This square mass in a great measure destroys the effect that Nature has attempted so successfully to produce, and is one among very many unfortunate instances of harm done to picturesque scenery through a lack of a little architectural knowledge in a rural way. The practical advantage gained is at once allowed; but the important point is, that an equally satisfactory and convenient result might have been attained, not only without any sacrifice of picturesque effect, but with even some advantage in this respect. For example, the boarded structure might have been omitted altogether, and a rough stone wall built up in great blocks, and without mortar, to the requisite height in a bold, irregular manner, could have received a platform at the required level, and a small, picturesque building might have been placed on this platform, if needed. In a year or so, by this arrangement, the rude new wall would have been covered up with vines and creepers, so that it would have added an actual twenty feet to the real height at this point, and an apparent height, when seen from below, of thirty or forty feet. As the building stands at present it dwarfs the appearance of the fall, and can hardly fail to strike the eye as a decided blemish, obtruded on the attention at the most interesting point of view in the whole landscape in which it occurs.

DESIGN FOR AN ARTIST'S STUDIO.

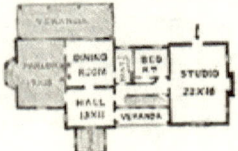

PLAN OF PRINCIPAL FLOOR.

N. E. VIEW.
SHOWING THE COTTAGE COMPLETED.

DESIGN No. 8.

PERSPECTIVE VIEW.

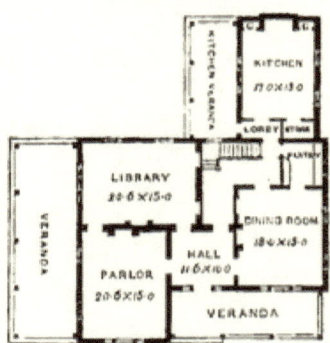

PLAN OF PRINCIPAL FLOOR.

DESIGN No. 8.

A SMALL COUNTRY HOUSE WITH KITCHEN WING.

This design was prepared and executed for Mr. R. L. Case, of Newburgh; and the general idea of the plan includes so much that is called for by the American climate and habits of life in the Northern States, that it will probably be better worth the attention of those who wish to build a moderate-sized cheap house, with a kitchen above ground, than many other plans of more pretension. It possesses one convenient quality, which some other styles of plans can not be arranged to include, for it admits of many modifications, without sacrificing its advantages. It may be completely altered in outside appearance, and doubled in extent of interior accommodation, and yet be in reality the same plan. It can be adapted to almost any situation by a proper arrangement of the roofs. Thus, for example, on an elevated and somewhat open site, such as the one that Mr. Case's house occupies, a

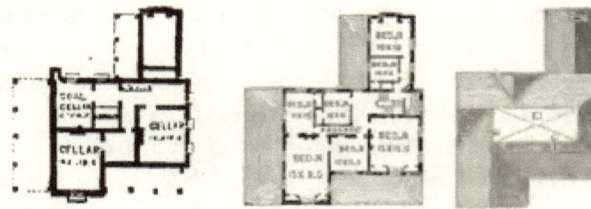

roof of only moderate pitch is desirable. On level ground, or in a valley, a high-pitched roof should be preferred. It is also an economical plan for the accommodation afforded, as will be seen by the particulars or cost that are annexed. The house, as now finished, is constructed with an eight-inch brick wall, furred off outside, and covered with clap-boards in the ordinary way followed in a wooden building. This plan of construction was adopted in accordance with the special request of the proprietor, who preferred it to any other method. Its advantages are, that it secures to a certainty a perfectly dry interior wall. On the other hand, it seems undesirable to have a brick house, and to give it the appearance of a wooden one, as brick is the superior and more durable-looking material. The accommodation may be thus described: A veranda-porch on the east provides a covered approach to the front door. The principal hall, 11 6 × 10 feet, gives access to the parlor and library, both of which are on the south of the house, and also to the dining-room. Another door opens on to a staircase-hall, which is easily accessible either from the principal rooms or from the kitchen wing. This is desirable, as the scale of the house would not warrant a second staircase. An east and a south veranda are supplied to the principal rooms, but each has windows that are unobstructed by any veranda. The dining-room connects through a pantry with the kitchen wing, which is also approached from the main body of the house under the staircase. A lobby opens on to a kitchen veranda, facing south, that provides a servants' entrance, and is convenient for hanging out clothes under cover in rainy weather. A kitchen, 17 × 13, fitted up with closets, wash-trays, and store-room, completes the accommodation on the main floor and wing. By this plan the disadvantages of living in the basement are entirely avoided, and the lady of the house can superintend her servants with ease and comfort.

In the chamber plan will be found five bedrooms, and a bath-room and water-closet; and in the wing two bedrooms, a linen-press, and a house-maid's sink. All these rooms are supplied with registers near the ceiling that communicate with foul air flues separate from the chimney-flues. In the garret over the bath-room is a large, well-lighted linen-room; and as this is planned on the half-landing, it is very easy of access from the chamber floor. A large store-room, the size of the bedroom over the dining-room, is finished off under the roof in a common way, and is secured with a door after being inclosed from the stairs by a plastered partition. The remainder of the space is open and unplastered. It makes a very roomy garret, with plenty of headway all over it; but the windows in the peaks are, of course, close to the floor, and it was never intended that any bedrooms should be fitted up here. The roof is covered with shingles, the flat being floored and covered with canvas. In the basement are cellars and furnace-room, the kitchen-wing foundations not being carried down farther than was necessary to keep clear of frost. In this house special precaution was taken, by Mr. Case's request, with regard to the plumber's work. All the pipes, hot, cold, and waste, were inclosed in a tin envelope fitted tolerably closely to the pipes. As the work proceeded this tin case was soldered up every here and there, and particularly where the pipe is led through the wall, in the first instance, and where it starts from the boiler. By this means the little insects that work their way from below, and are often found about water fixtures in rooms, are prevented from crawling up and down, and breeding among the warm pipes, as they are tempted to do in many situations.

The carpenter's contract for this house was taken at $3500; the mason's at $2500; the remainder of the work was done by the day.

After the contracts had been made the proprietor left the work entirely in the hands of the architect; and, with the exception that hard walls were substituted for brown walls throughout, and that some trifling alterations were made in the arrangements for the linen-press, the plans, as signed, were faithfully executed for the contract amount, without any difficulty whatever. The carpenter's and mason's extras, which amounted to $350, included the change from brown wall to hard finish, and all the work appertaining to a large outbuilding at some distance from the house.

The vignette illustrates an unexecuted study for a garden outbuilding, and is supposed to be located in a situation where it *must* be seen more or less.

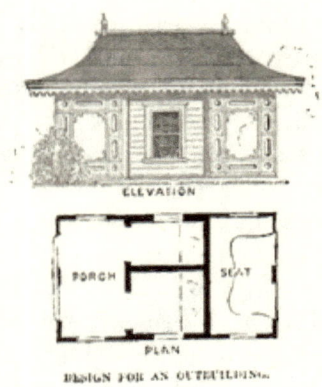

DESIGN FOR AN OUTBUILDING.

DESIGN No. 9.

PERSPECTIVE VIEW.

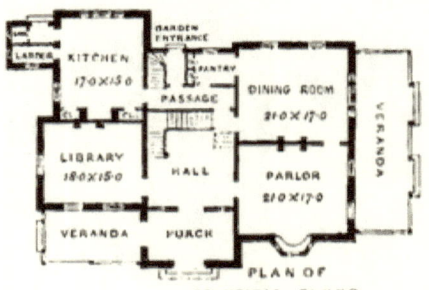

PLAN OF PRINCIPAL FLOOR.

REAR VIEW.

DESIGN No. 9.

AN IRREGULAR BRICK COUNTRY HOUSE.

In this design, which was made for Mr. J. Robins, of Yonkers, a brick porch, connected with a veranda, and so arranged that it can be inclosed in winter, leads to a roomy, well-lighted hall, in which is an open staircase to the chamber floor. A staircase of this sort may be made quite an agreeable feature in an open front hall, if there is a back stairway to the bedrooms. This is provided in the present case, so that the servants need not use the principal stairway when attending to the chambers. The library is a pleasant room, 15 × 18, unconnected with any other apartment, and with windows opening on to the veranda. It is entered from one side of the hall, and the parlor, 17 × 21, from a door opposite. This room is supplied with a bay-window, and connects with the dining-room, which also communicates directly with the main hall, and with a pantry, or service-room, that has a private ac-

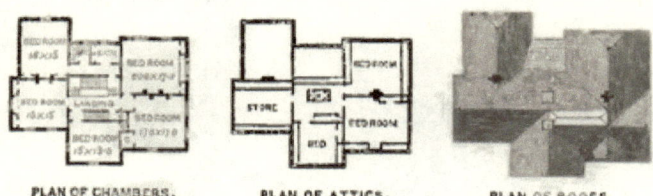

PLAN OF CHAMBERS.   PLAN OF ATTICS.   PLAN OF ROOFS.

cess from the kitchen department. Both the parlor and dining-room open on to a wide veranda. The kitchen is 15 × 17, and has an entrance-porch, with sink and larder. The lower floor is occupied for cellars and furnace-room, and is not finished off. In the plan of second floor will be found five roomy chambers, a bath-room, and a linen-press under attic stairs. One of these, which would probably be a spare room, occurs over the kitchen, and has a ceiling following the line of roof and collar beams, which makes it more airy and agreeable, and improves its appearance, without sacrifice of space, as there are no attic rooms over the kitchen department. In the attic of main building three bedrooms are finished off. It also provides a large garret, and a skylight to main stairway. The plan of the roof is deserving of some attention, as it covers a large space, without occupying much room in the flat. The rear of this house looks down a somewhat steep hill, at the foot of which runs the principal approach road; and the roofs are, therefore, hipped back on this front, so as to prevent them from appearing in too acute perspective when viewed from below, as would otherwise undoubtedly be the case.

This house is built at Yonkers, of brick, the ventilator and dormer-window shown on the drawing being subsequent additions.

BOAT-LANDING IN THE CENTRAL PARK.

DESIGN No. 10.—(V. & W.)

PERSPECTIVE VIEW.

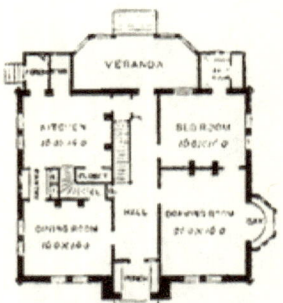

PLAN OF PRINCIPAL FLOOR.

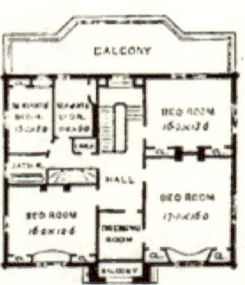

PLAN OF CHAMBERS.

DESIGN No. 10.
(V. & W.)

SUBURBAN HOUSE WITH ATTICS.

This is a study for a simple suburban brick house of moderate size, square on plan, with the exception of a small central projection; and it depends almost entirely on the roof lines for any picturesqueness of character that it may lay claim to. It has been designed and executed for a gentleman residing in Newburgh. The plan has been slightly modified in execution, the bedroom on the principal floor being finally arranged as a second parlor; but as this does not improve the plan for our present purpose as an example of peculiar arrangement of accommodation, the original design is adhered to in the illustrations.

The house is entered by a recessed porch, with a covered balcony overhead. This upper balcony being also recessed in the brick-work, and inclosed at the sides, is always in shadow, and materially helps to relieve what would otherwise be a monotonous front. This space is fitted with a glazed frame in the winter, the porch being inclosed with folding-doors as soon as the cold weather sets in. The hall extends through the house, and communicates with a parlor and bedroom, a dining-room, and a veranda in the rear. The main stairs are in this open hall, and on the half-landing is a connection, through a French casement-window, with a balcony over the veranda, from which a pretty view is gained. The parlor has a large bay-window in it, the cornice of the room being carried round the recess that it forms. There is a private door from this room to the bedroom adjoining.

It is not generally a desirable plan to give up the space necessary for a bedroom on the principal floor; but circumstances occur in which it is a very great desideratum, and this study may serve to show how, in a simple house, the idea may be developed. It will be perceived that a portion of the veranda is inclosed for a small dressing-room to this bedroom, thus making it a far more commodious sleeping apartment than it would otherwise be. The dining-room connects with a pantry, and is also supplied with a large china-closet. The pantry is fitted up with hanging-shelf, drawers, and closet, and connects with the kitchen, which is thus shut off from the living-rooms, although under the same roof as the rest of the house. An inclosure of the veranda, similar in size to that on the opposite side, supplies a space for a pantry and sink-room. The servants' entrance is quite convenient of access from the road, but, at the same time, is shut off by its position from interfering with the privacy of the veranda. A door, where shown near the hall door to veranda, incloses the basement stairs for the use of the kitchen, and a compact flight of stairs from the kitchen itself provides a separate access for the servants to the bedrooms above. This staircase occupies a very small space, and is a great addition to the convenience of the house. In the basement is a wash-room under kitchen, with an outer entrance, close by servants' entrance, for convenience in carrying out clothes to dry. The remainder of the space is not finished off, and furnishes cellars and furnace-room.

A straight veranda inclosed on both sides would

not, perhaps, be thought sufficiently airy, and a projection is therefore made, as will be seen on reference to the plan, to increase its size and give it a more open effect. This arrangement also adds somewhat to the external appearance of the design, at but little additional expense, while it is calculated to insure privacy in a suburban house; and in common houses the notion is carried out frequently, in a simple way, by lathing up the ends of verandas, to prevent them from being overlooked by next-door neighbors. Such a veranda as is here shown will be almost as retired as any of the rooms inside the house.

Up stairs are three full-sized bedrooms, and a small bedroom, or dressing-room, a bath-room, and water-closet, a linen-press, and two servants' bedrooms, the latter disconnected with the other apartments. This arrangement is made with the idea that the attics are to be left entirely unfinished for a time, the house being occupied by a small family; but the plan has been, from the first, so arranged that three or four airy, well-lighted rooms can be fitted up at any future time, and if this should ever be done, the two servants' rooms shown on chamber plan might be converted to the use of the family, and the servants' rooms arranged above.

In the actual execution of this design the owner determined, during the progress of the work, to throw the two small bedrooms into one, and to finish off a portion of the attic for the use of the servants.

It seems desirable, in planning a country house, to locate the principal rooms in such a manner that they may court the pleasant southerly breezes in summer, and the southerly and easterly sun in winter. In some situations it so happens that this can not be done, except at an entire sacrifice of the pleasantest views, and the general idea of the plan in such cases may require to be modified accordingly; but under ordinary circumstances an architect will so manage matters that the inferior rooms and offices will quietly slide into the uninviting north or northwest corner of the house, and thus occupy that portion of it which can best be spared from the living-rooms. There is, however, a disadvantage in this natural arrangement which, if not guarded against, may give trouble, for the water-fixtures, being on the north side of the house, may be affected by a severe frost; but this may be avoided in several ways. One plan is to keep all the pipes on the inside partitions of the house, and to protect them from any exposure by casing round them, and filling in with sawdust or other non-conducting material. This is all that can be done in a house without a furnace, except to carry the kitchen flue through the cistern; but where a furnace is used the remedy is easy and complete. A two-inch tin pipe communicating with the hot-air chamber should be carried up with the water-pipes, and after being continued through the bottom of the cistern, should be coiled once round it on the inside, and then be left with an open outlet just above the level of the overflow pipe. If the cistern is afterward covered with a partially air-tight lid the result will be found entirely satisfactory, as it has proved in the house illustrated on page 180, and in others over which I have had the control.

The roof of this house is covered with the purple and green slate from the Vermont quarries, arranged in stripes, as shown, and the effect, as executed, is very soft and agreeable. The cost was scarcely more than for a shingle roof of good quality, the slate being provided and put on with mortared joints at $7 50 a square. Circumstances, however, were favorable in this instance, and $8 per square is as little as this slate can be generally laid for.

It will be seen that a special arrangement is made for the carrying off of the rain-water, the pipes being fitted with ornamental heads, and connected with the gutters by brackets brought down on to the brick piers at the angles. Some additional individuality is thus given to the external appearance of the house, and attention is requested to this part of the design, as it refers to a point that is very generally left unstudied, many designs being, in consequence, marred by an awkward, obtrusive arrangement, or rather want of arrangement, of the rain-water pipes.

The cost of this house, including the fence shown in the vignette, and a moderate-sized stable and coach-house, was as follows:

Mason's work	$1921 02
Carpenter's work	3278 71
Stone-cutter	380 00
Furnace	201 50
Gas-fitting	85 54
Plumbing and tinning	945 49
Painting	641 32
Bell-hanging	46 40
Architect's commission (5 per cent.)	375 00
Total	$7874 98

A recommendation by an architect to his client to build a larger house than he or his family require for comfort, would seem, at the first glance, to be invariably bad advice to give to a prudent person; but experience shows that it is sometimes the best course to pursue. If, for example, a gentleman, whose requirements are comparatively small, purchases a valuable lot in the neighborhood of a thriving country town, he will certainly damage the selling value of his property by erecting on it a house that can not be easily converted into a tolerably roomy family residence. It is, therefore, in such a case, a better investment for him to expend somewhat more at first, and arrange his design so that further accommodation can, without much trouble, be obtained, than to cut and pare his house down to the exact measure of his own immediate needs, without reference to its probable market value. Some men spend, while others spare, extravagantly, and either habit is found to be inconveniently unprofitable in the long run.

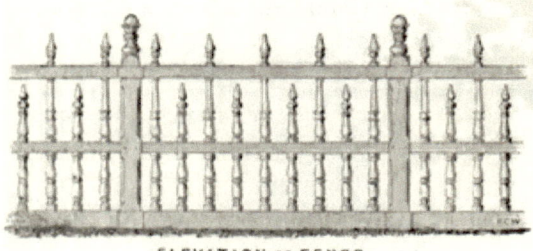

ELEVATION OF FENCE.

DESIGN No. 11.

A NEARLY SQUARE SUBURBAN HOUSE.

This design has been executed in brick by a gentleman in Rondout, on a site overlooking an ample stretch of the Hudson River. It contains about the same amount of room as the design last described, but

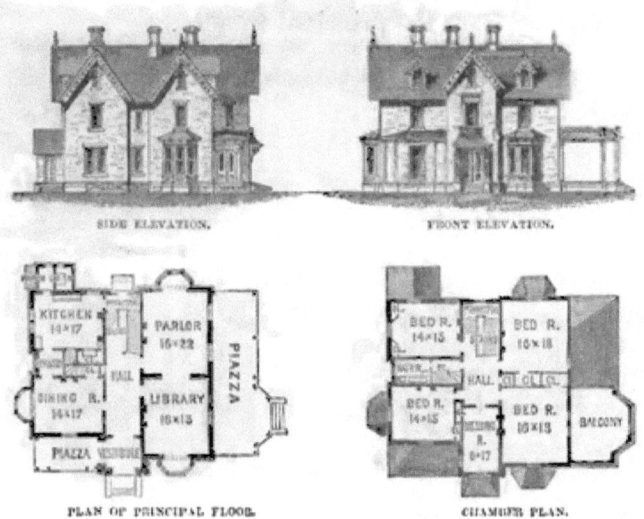

arranged in a somewhat different manner. The exterior outline is more picturesquely broken, and the whole design is on a somewhat more extensive scale. The smaller piazza is so arranged that it can be readily converted into a plant cabinet in winter, and a bay window is introduced into each of the principal rooms.

The cost of this house, including painting and plumbing work, and a coach-house and stable of liberal size, was about ten thousand dollars in 1862.

The vignette illustrates a design carried into execution in the neighborhood of New Haven, Connecticut. The special requirement in this case was the introduction of a parlor much larger than the dining-room or library, into a square house that should have the kitchen under the same roof with the rest of the building, but shut off from the principal rooms.

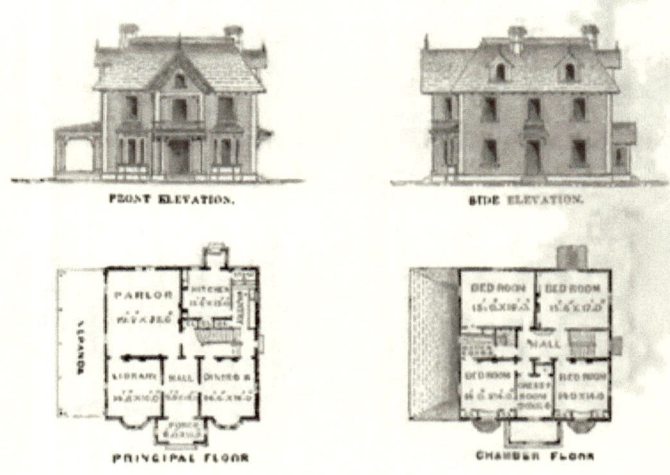

DESIGN No. 12.

AN IRREGULAR HOUSE, WITHOUT KITCHEN WING.

This design, which is in process of execution, was prepared for a somewhat peculiar site in the vicinity of Springfield, Mass. The ground descends rapidly to the south, and the entrance to the house, on account of the shape of the property and the situation of the approach-road, was of necessity made on the north front.

Attic bedrooms throughout were not desired, but

PERSPECTIVE VIEW.

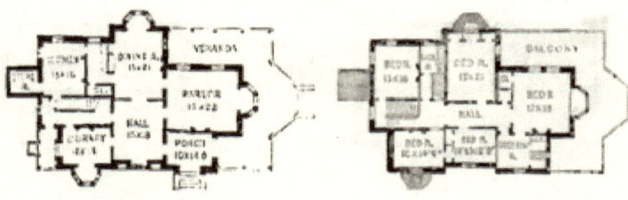

PLAN OF PRINCIPAL FLOOR.   CHAMBER PLAN.

190   VILLAS AND COTTAGES.

some accommodation for servants was needed over the bedroom floor.

The main body of the house is only two stories high, but the block inclosing the kitchen is really four stories high, as is shown in the engraving.

The vignette illustrates a house erected in Llewellyn Park, in the neighborhood of Orange, New Jersey.

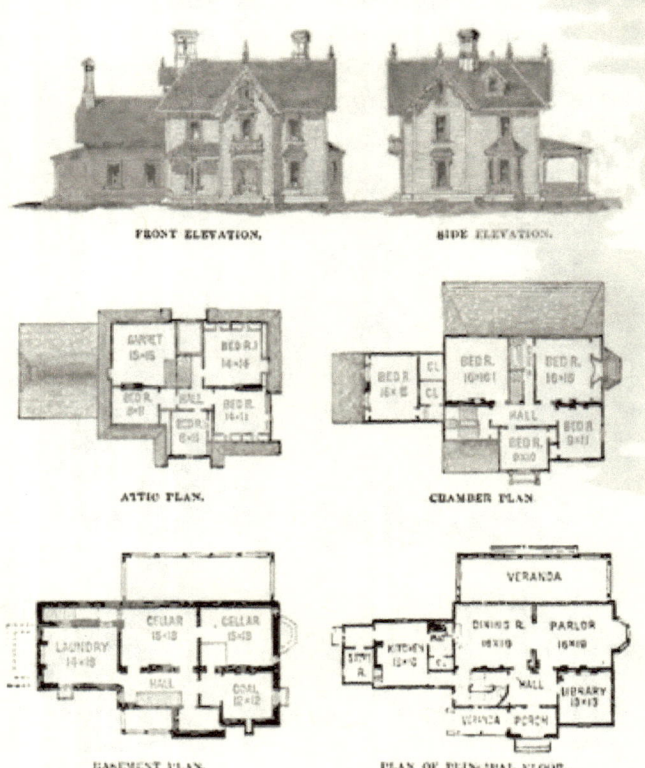

DESIGN No. 13.

PERSPECTIVE VIEW.

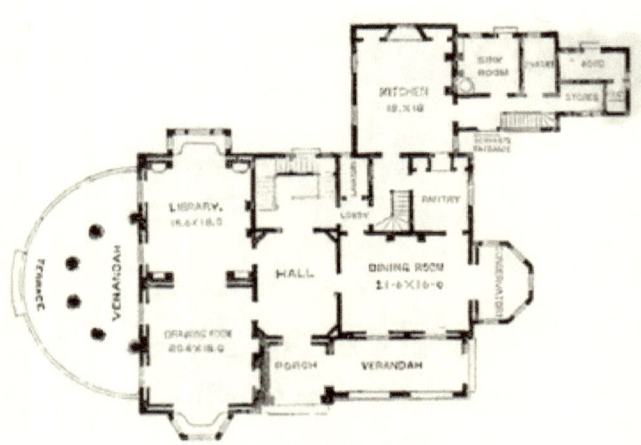

PLAN OF PRINCIPAL FLOOR.

DESIGN No. 13.

A WOODEN VILLA WITH TOWER, AND WITHOUT ATTICS.

This design was prepared and executed for Mr. C. H. Rogers, at Ravenswood, Long Island, and as it is built on a situation which commands a good view of the East River, more or less intercepted from the lower stories by buildings and trees, it was thought desirable to arrange some point of view from the upper part of the house which should admit of the whole extent of prospect being conveniently seen. A tower three stories high, finished above with an octagonal observatory, easily accessible from below, has, therefore, been included in the design, as will be seen on the sketch. By this plan an interesting view is obtained, clear of the roofs in every direction. The observatory was made of an octagonal form, with a projecting balcony round it, so as to reduce its per-

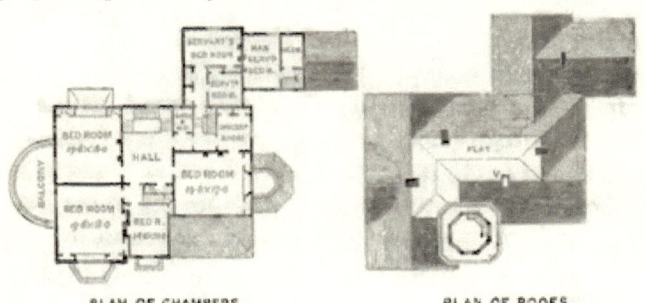

PLAN OF CHAMBERS.   PLAN OF ROOFS.

spective dimensions, and give a more light and airy effect to the upper part of the design; for if a square tower four full stories in height, even though it may be quite small on plan, is grouped in connection with a villa of but moderate size, the probabilities are that it will appear to have an undue preponderance over the other features, and will thus compel the more strictly domestic parts of the house to seem of secondary importance. This result is, of course, undesirable; for although the observatory room is intended to be both useful and ornamental, it is far less necessary to the real comfort and enjoyment of the inmates than the living rooms that are in occupation all the time. The lower story of the tower is occupied as a front porch, which is connected with a veranda, and communicates with an entrance hall. The angles of this hall are finished octagonally, so as to improve its general effect. The drawing-room is 18 × 20:6, and is finished with a large bay-window at one end. It also communicates, through side windows, with a semicircular veranda and a balustraded terrace on the river front. The library opens from this room and from the hall. It is furnished with recessed book-cases and a square bay, and, like the drawing-room, communicates with the veranda and terrace. The dining-room on the other side of the hall instead of a bay-window has a conservatory, or plant cabinet, attached to it, as will be seen on the plan. It is also provided with a roomy pantry that has closets and a sliding hatchway immediately connecting with the kitchen. The main staircase is in a second hall, and under the upper landing is a large cloak-closet and an entrance to the garden. A little dressing, or lavatory room, is planned near to this hall, and the servants' staircase and passage-way to the offices connects with the main building at this point. The kitchen is in a wing on the same level as the other part of the house, and has a sink-room, pantry, and wood-house attached. It also contains a small separate stairway to two upper rooms for men-servants.

This house was built of wood filled in with brick, the principal rooms being provided with sliding-shutters; and the carpenter's and mason's work for this building, and a coach-house, etc., including painting, was done for about $10,000, which, with architect's commission at five per cent., amounted to $10,500.

WEST ELEVATION

An illustration of the river front is shown above, drawn to a somewhat larger scale than the perspective view, so that the arrangement of gable and mode of finish may be intelligible by a reference to the plan of principal floor on previous page. The formal square appearance it presents, compared with the actual effect of the villa in execution, or the other illustration of the same house already given, will serve to show how incomplete an idea is likely to be formed of an architect's design, if it is judged of from elevation only, and yet it is by no means uncommon to find this mode of illustration adopted in architectural works in preference to any other. If a study for a house is proposed to be so drawn out that it may be used by builders for working purposes, it is absolutely necessary that plans, elevations, and sections should be furnished, because no measurements can be taken from a perspective drawing, however neatly it may be done; but if the study is submitted with a view to show what sort of artistic effect may be produced, in execution, from a certain arrangement of ground plan, nothing but a perspective view will convey an accurate idea to the mind; and as the intention in this volume has been to give suggestions, rather than to supply cut-and-dried designs, the perspective mode of illustration has been adhered to as much as possible.

The vignette shows a design for a fence and gate executed in connection with the above design.

DESIGN No. 14.—(D. & V.)

PERSPECTIVE VIEW.

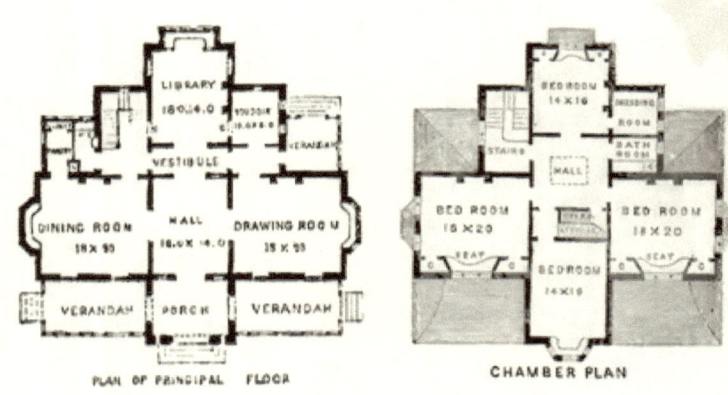

PLAN OF PRINCIPAL FLOOR.   CHAMBER PLAN.

DESIGN No. 14.

(D. & V.)

A SYMMETRICAL COUNTRY HOUSE.

This design was prepared and executed in the vicinity of Newburgh, with the exception of the ventilating turret, which was a suggestion offered at the time the building was erected. This prominent feature has not, however, yet been carried into execution, as the gentleman for whom the plan was prepared preferred to omit it; still, it forms an integral part of the design, and is introduced in the sketch, as the composition appears to be somewhat incompletely represented without it. There were some special requirements made by the proprietor in this instance that are, perhaps, with some readers, calculated to give additional interest to this plan. The house during the summer months was to be, to all intents and purposes, a Southern house; ample circulation of air

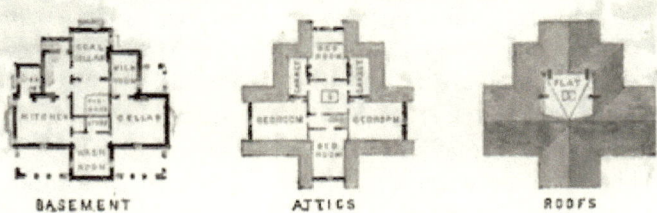

BASEMENT.   ATTICS.   ROOFS.

was to be provided, with plenty of veranda space; and a cool, open arrangement of rooms was especially asked for, as the house would be required to accommodate agreeably many residents and visitors during the hot weather. On the other hand, the plan was to be so arranged that during the colder part of the year the house should be suited to the changed needs of its fewer inmates, for it was the intention of the proprietor to occupy his residence during the winter. A completely fitted up and somewhat decorative principal floor was also a point of importance with the proprietor, who did not propose to expend any large amount on his house, and whose instructions were for rooms of but moderate size, and for a basement kitchen, so as to avoid the extra expense of a kitchen wing. In the plan it will be observed that a brick porch, connecting two verandas, and forming with them one continuous piazza, opens on to the principal hall, which is 14 × 18. This hall communicates, by sliding doors, with the drawing-room on one side, and the dining-room on the other. Each of these rooms has a bay-window at the end opposite the hall entrance, and thus, when the doors are thrown open in summer, an agreeable vista effect is produced, and a free circulation of air is provided for. The upper end of the hall is traversed by a light, ornamental, open, arched screen, which is introduced so as to give a definite character to a passage-way, or vestibule, as shown on the plan. Beyond this arched screen is an entrance, with sliding-doors, to the library, and at the opposite end of the latter room is a large square bay-window, with open screen-work and seat in connection with it. Thus another extensive vista is obtained in summer evenings through the house in this direction, and when the doors are open, any one sitting in the library bay can see the river view framed, as it were, in the outer arch of the porch. A little boudoir, or ladies' morning-room, with a private veranda entirely shaded during the whole of the forenoon, is planned close to this library and to the drawing-room, and a similar space is inclosed on the opposite side of the house for water-closet and pantry, which has a lift in connection with the floor below. It will thus be seen, by an inspection of the plan, that it would hardly be possible to have a more airy and open arrangement for summer; for, standing in the hall when the rooms are thrown open, one can see clear through the house, north, south, east, and west; and the porch, hall, vestibule, library, dining-room, drawing-room, and veranda, are converted, as it were, into one connected apartment. All idea of the moderate size of the single rooms, the largest of which is but 18×20, is thus done away with, and the house necessarily seems roomy, open, and ample in its accommodation.

So much for the summer arrangement. In winter, on the other hand, the first step should be to inclose the arches of the porch with glazed frames, and the next to close the sliding doors for the season, entering in future each of the separate rooms from the ordinary-sized doors which are provided to each of the principal apartments for this purpose; the library being approached through a book-case door, already illustrated and described in the opening chapter (see page 90). The furnace may then be started, and the house will be found to be a warm winter house, suited to a severe climate. All the thorough draughts are shut off, and the separate rooms are small, readily warmed, and easy of access from the chambers. The bedroom floor in this house contains an upper hall, lighted by a skylight, four full-sized bedrooms, a dressing-room, and a bath-room with water-closet. The attic contains two spare bedrooms, shut off entirely from the apartments for domestics on the same floor, and it also provides an open hall, two servants' bedrooms, a garret, and lumber room.

In the basement will be found the kitchen, washroom, milk-room, furnace-room, and cellars. The carpenter's and mason's contracts for this house were taken at $7230, and the painter's, and plumber's, and decorator's accounts, with some ornamental ceilings, and other carpenter's work inside the house, not contemplated in the original contract, made the amount expended, and on which five per cent., architect's commission, was charged, $9326 51. The stable, which contained accommodation for three horses, a coach-house, a harness-room, a coachman's living room, with bedroom over, and a hayloft in the roof, was built of brick for $1700. It is illustrated in the vignette below.

DESIGN FOR STABLE, ETC.

DESIGN No. 15.—(V. & W.)

PERSPECTIVE VIEW.

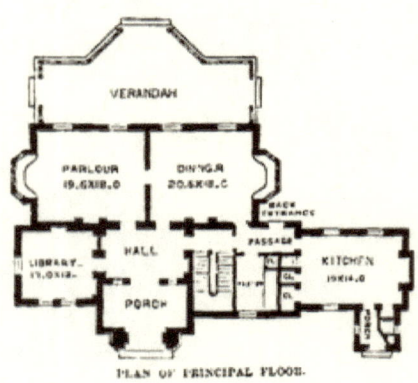

PLAN OF PRINCIPAL FLOOR.

DESIGN No. 15.
(V. & W.)
BRICK VILLA WITH TOWER, AND WITHOUT ATTICS.

The working plan of this design has been prepared for Mr. Walker Fowler, of New York, and the house is designed to be erected by him on an agreeable site at Limestone Hill, in the neighborhood of Newburgh. The location has much individual character, and the whole property is well adapted for a country seat. Such a spot, however, is scarcely appreciated in the vicinity of the Hudson River, for an extended water view is, under such circumstances, thought an all-important consideration; consequently a situation like this, in which undulating hills intercept the river view from the lower rooms, appears to much less advantage than it would if it happened to be in another

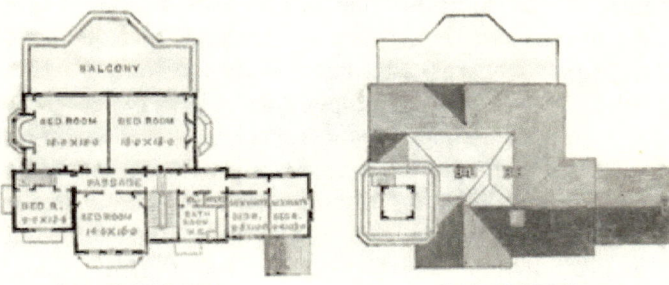

PLAN OF CHAMBERS.   PLAN OF ROOFS.

part of the country altogether. The site for the house is on an elevated space of table land. On the veranda front the hillside descends abruptly and picturesquely some sixty or eighty feet within full view of the drive road, and as this steep hillside is well covered with healthy evergreens and fine deciduous trees, it will form, as it were, a broad and handsome base for the house, part of which, especially the observatory, would be visible from below. On the other side the formation of the ground is such that the house will be approached without any sudden ascent, as Nature has kindly made ample preparation within the boundary line of the property for an easy winding road up to the top of the hill. On the south side is a stretch of fine woods, on the edge of which is just the spot for a stable, and scattered here and there among the trees are large boulders of stone, grouped, by accident, three and four together, quite artistically, and forming pleasant natural seats and nestling-places for ferns and wild flowers. One mammoth boulder, that looks like a sleeping monster, some twenty or thirty feet long, has been split in a peculiar manner by a tree growing into a fissure beneath it, and thus presents a very curious appearance. Large stones like this are often found in quite unaccountable places, and it is supposed that they must have been borne to the spot, and dropped, by melting icebergs, ages ago. This solution, whether true or not, is certainly ingenious. But to return to our more immediate business, which is a description of Mr. Fowler's plan. A recessed porch, large enough to serve as a morning veranda, provides access to the principal hall, which is only of small dimensions, but it is provided with cloak-closets, and contains the doors to the three principal apartments, and to the staircase. The library is in the lower story of the tower, and the design is so arranged that this tower can be omitted entirely when the first contract is made, without a disagreeable appearance being the result; and although the interior accommodation and the exterior effect of the house would be materially reduced under those circumstances, the design would still be fair and complete, and the addition could be made at any time. A parlor and dining-room open on to a veranda. Near the dining-room is a pantry, a garden entrance, and door to the kitchen, which is in a wing building.

The chamber plan supplies four bedrooms, and a fifth in the upper story of the tower, also a bath-room and water-closet, a linen-press, and two servants' bedrooms. The observatory is conveniently reached by continuing the staircase that leads to tower bedroom. The roof is arranged as shown on the plan. The intention in this design is to insure, as far as possible, an irregular picturesque effect, without any sacrifice of convenience or a large outlay of money. As the house is to be built on somewhat of a highland, it seems undesirable to use an acute pitch for the roof, for the trees that surround the site proposed for the house, although vigorous and well shaped, are somewhat scattered, and of no great magnitude. They would, therefore, scarcely take their proper share in the general composition, if the roof were made too prominent a feature. In designs like that for Mr. Willis's house, on the other hand, the rear view is so enveloped with hardy evergreens, omitted for the most part in the drawings, that it is preferable to use a high-pitched roof, as the house would otherwise be in a few years *entirely* shut out from view. Considerable judgment is needed in settling on the exact position for a house like this, so as to realize all the advantages that the site affords. It must not seem to overhang the descent, or the effect will be crowded, and will give the idea from the road of a small, restricted property. Neither should it retreat very far from the brow of the hill, or the house will be shut out of sight, and altogether lost on a tolerably near approach to the premises. A happy medium, both in the location of the site and in the pitch of the roof, is the desirable point to aim at under such circumstances.

The vignette shows a design for an observatory to be erected at Hillside Cemetery, Middletown, N. Y.

DESIGN No. 16.

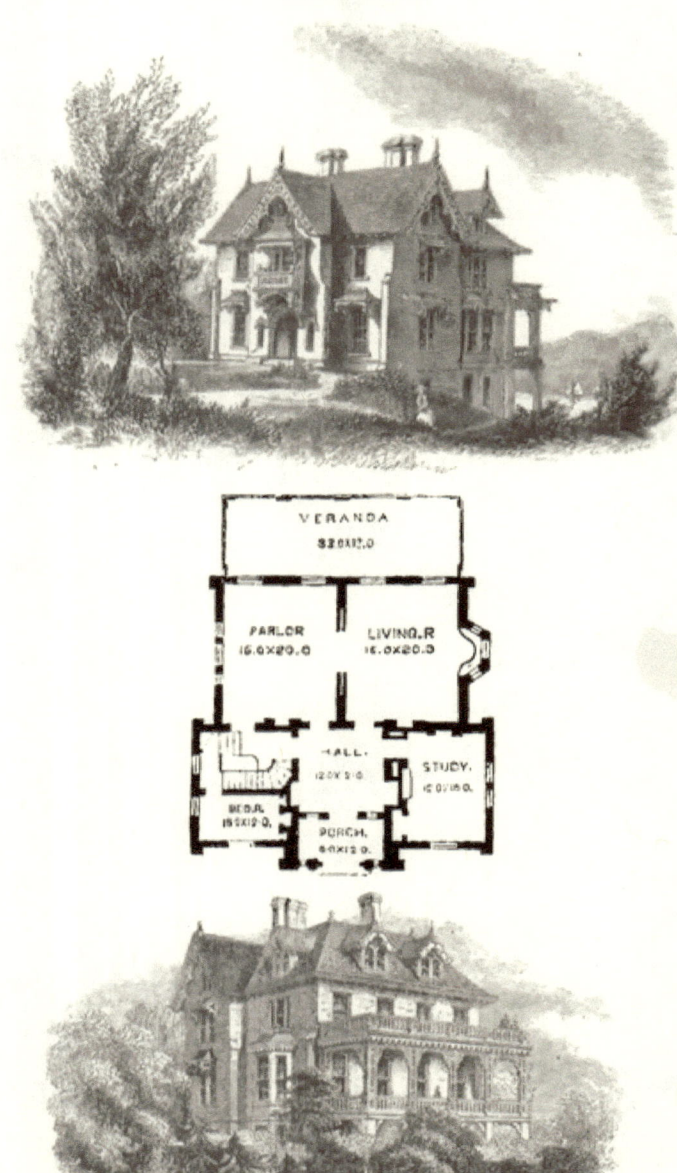

DESIGN No. 16.

A PICTURESQUE SYMMETRICAL HOUSE.

This design was prepared and executed for Mr. W. E. Warren, of Newburgh. It is situated in Montgomery Street, which is a straight road running parallel with the river, at a considerable elevation, commanding beautiful views of West Point, the Highlands, and the Hudson. From this level the cross-streets descend with a steep pitch to the shore, and the building sites, in consequence, slope considerably.

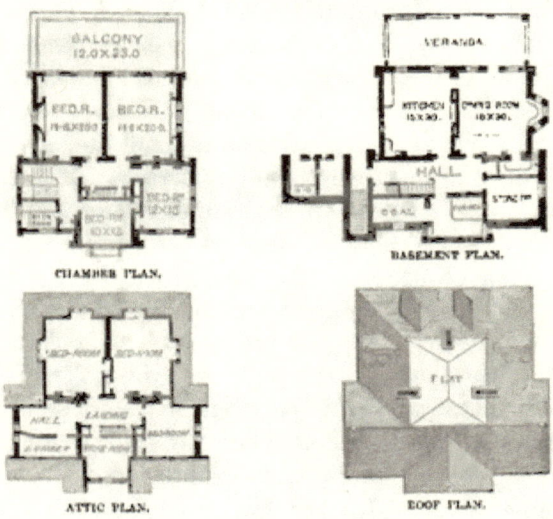

As Mr. Warren's house was to be built on a corner lot, this peculiarity of position became very conspicuous, and required to be considered with some care, when preparing the plans, so as to avoid a stilted, disagreeable effect. The entrance front of the house faces on Montgomery Street, and is seen in the ordinary way on a level with the street. The design is therefore made with gables accordingly. The rear, on the other hand, is conspicuous, chiefly from the lower streets and from the river; and as it provides a basement story entirely out of ground, on account of the rapid descent of the hill, the whole front is, in consequence, of an altered proportion, and required a different treatment. The roof in this part of the design is hipped back, without gables, as will be seen on the lower view, and dormer-windows are introduced to give light to the attic bedrooms. The apparent height of the rear elevation is materially reduced by this arrangement, and the general effect is altogether made more easy than if gables had been introduced corresponding with those on the opposite side.

A recessed porch opens into a hall twelve feet square, in which are cloak-closets, and the doors to the principal apartments and staircase. The room called study is fitted up with wash-stand, etc., in a closet, so that it may be used as a bedroom or sick-room at any time; and the small room by the staircase is well-suited, and is at present used for a private office instead of a sleeping apartment, as at first proposed. The doors in the hall are grouped together, with a panel for a picture between each pair, as shown and described on page 88. There are two parlors connected by sliding doors, and opening on to a piazza twelve feet wide. This last feature of the composition is made very roomy, as it commands an extended view of the river, while the ornamental plot of ground attached to the house is but of small dimensions, and is at a much lower level. This veranda, therefore, is sure to be the chief summer resort of the inmates, and is made of large size, with a balcony over, which, as it faces east, is in shadow during the evening, and affords an agreeable accessory to the bedrooms connecting with it.

The chamber plan contains four bedrooms, bathroom, water-closet, and linen-press. The attic provides several spare rooms, as shown. In the basement is the dining-room, with a door on to the veranda, and also a roomy kitchen, with other offices. The necessary shown on this plan is arranged as described on page 47.

This house is built with an eight-inch wall throughout, above the level of water-table, and the beams are supported on iron rests, as described in the opening chapter. The interior has been well finished throughout under the superintendence of the architect, and the work was done by the day, in the best manner, and of the best materials, the cost being as follows; viz.,

Carpenter's bill	$2,191 08
Lumber	1,428 58
Doors	237 45
Mouldings	40 99
Bell-hanging	45 00
Glazing and canvas	285 26
Mason's bill	2,266 45
Brick	426 15
Cut stone and paving	940 63
Cartage	88 50
Iron castings	54 23
Tinner's work and plumbing	1,184 37
Gas-fitting	75 21
Painting	895 23
Furnace	204 78
Mantles	614 50
Architect's commission (5 per cent.)	569 88
Total	$11,967 60

The mason's bill in a contract often covers the tinner's work in connection with leaders, which is not here charged to him, as that and the plumbing were in one account. The bill for paving, on the other hand, includes considerable work that would scarcely belong to a mason's contract. The cost of grates and range is not taken into the account.

The particulars furnished above may be of some interest to any one about building such a house, as they show the relative amounts of the various bills; but there is so much variety in the style and consequent expense of interior comforts, appliances, and finish, that a detailed estimate of this sort is not so valuable as might at first be supposed.

The vignette shows a plan adopted for laying out the grounds. By this plan two entrances are planned on Montgomery Street, and one on the side-hill, while a space large enough for a cart to get to the garden with manure, etc., is arranged on the left hand side. A piece of ground of this size should not at any time be filled with very large trees, for flowering shrubs, evergreens, and a few fruit trees would be all sufficient.

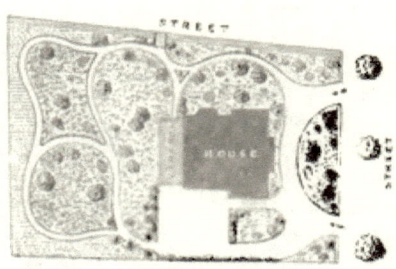

PLAN of GROUNDS.

DESIGN No. 17.

SOUTH-EAST VIEW AS ALTERED.

SOUTH-EAST VIEW BEFORE ALTERATION.

NORTH-WEST VIEW AS ALTERED.

DESIGN No. 17.

ALTERATION OF OLD HOUSE.

This design, which is for alterations and additions to the residence of Mr. Thomas Powell, of Newburgh, was planned some few years ago from the instructions of his son-in-law, Mr. Ramsdell, the president of that gigantic American fact, the Erie Railroad, and it is selected for illustrating this part of our subject from several studies for alterations, executed from time to time in the same neighborhood and elsewhere, because it seems to show with clearness the leading points that require attention in such improvements. There are to be found in different parts of the country many families who have been settled for several generations on the same spot, and their old, simple wooden homesteads, mended and patched every few years, hold their own with commendable pertinacity. They have no idea of falling to pieces, and are altogether too solid and substantial to be pulled down. Now this quality of durability is, of course, in the abstract, an excellent virtue for a house to possess; but it must be confessed that, in such very awkward and ungainly structures as often fall to the lot of these well-settled families, its presence could be cheerfully dispensed with, were it not for the many interesting associations and family reminiscences that linger round the old house, which has been, perhaps, the home of the father's and grandfather's childhood. These associations are so valuable, and so little fostered by the ordinary course of events in American families, that they deserve to be cherished in every possible way; and it may therefore, now and then, be more wise to do the best that can be done with an old house, even at some sacrifice of external completeness of design and internal convenience, than to demolish it entirely and build anew from the foundations on a better plan. It does not often occur that a design can be altered so as to be entirely satisfactory; but much may be done to compromise matters without an uneconomical expenditure. The error generally fallen into in such cases may be thus described: Alterations are commenced without any very definite idea as to where they will end. One thing is done after another; partitions are pulled down, floors taken up, ceilings heightened, new windows and doors inserted, till the house is a complete labyrinth of mixed-up work, the clew to which is wholly undiscoverable by either proprietor or mechanic, and the natural result is, that, after many mistakes, and a severe trial of the patience of the owner, the workmen get through somehow, and are paid up and dismissed, while the house, although somewhat more convenient, is almost as ugly as before, and the proceeding, from first to last, has cost four or five times as much as the proprietor had an idea of laying out on it. This question should always be asked and answered fairly before commencing operations, viz., Is the house worth altering *at all?* Sometimes it is not. The frame may be decayed, the sills rotten, the floors out of level, the ceilings altogether too low for comfort; and as points like these can not generally be meddled with economically, it seems useless under such circumstances to spend much money on alterations and additions. If, on the contrary, the house is in a sound, substantial condition, and has no radical defect of interior arrangement that must always make it an objectionable residence, it becomes worth while to consider *how much* alteration and addition the house will bear profitably, and the whole plan and intention, from first to last, of the work to be done, needs to be studied and determined on beforehand more exactly and minutely than would be required for a new design altogether, for each part must be, as it were, dovetailed into the other, so as to get the advantage aimed at without awkwardness of appearance or undue sacrifice of the work already in existence. The fact is, that altogether *too much* is generally attempted. The best way is to do as little as possible beyond obtaining the leading features of arrangement and appearance that the alteration or addition is designed to procure. I remember once being called upon to pay a professional visit to a gentleman who wished to alter his house, which was a wooden one. I examined it, and found it had many serious defects, and advised him not to spend a cent upon it, but, if he was dissatisfied with the accommodation it afforded, to sell his present house and lot, purchase a fresh piece of ground, and start anew. He wanted to heighten all the ceilings to begin with, a process which would, of course, throw every door, and window, and beam out of position. Then the doors and windows must be made larger, and the frame must be new-sided, and the roof new-shingled, so that it became evident that what was really wanted was the old knife with the simple addition of a new handle and new blades. My advice seemed to be somewhat unsatisfactory to the proprietor, who evidently expected some encouragement, and perhaps an alteration was ultimately made; if so, I am certain that the result must have been even more unsatisfactory than the advice. In Mr. Powell's house the whole construction was in good preservation, and the addition shown on the sketch of the unaltered house had been executed substantially some years ago for the sake of the room it contained, without reference to its effect on the external appearance of the design; and the chimney of this wing being below the other roofs, a large cowl was required to prevent the chimney from smoking. All that was wanted in the way of interior enlargement was a study with bedroom over, which I planned in a square projection at the rear of the house, as shown on the lower view, and an enlargement of the parlor, which was arrived at by a square projection in front of the depth of the veranda, shown on the upper view, the old wall above the level of parlor ceiling being carried on iron suspension rods. Some of the windows and doors were shifted along a foot or two one way or the other, so as to bring the arrangement of openings into a form that would admit of proper treatment on the exterior. Small alterations, also, were made here and there to improve the internal convenience of the plan; but still nothing was done of sufficient magnitude to render it necessary for the family to leave the premises, even for a day, and the house was more or less occupied during the whole period required for the execution of the improvements. The chief alteration was made by taking a slice off the top of the original stiff old roof, and then bringing up the flat roof of the wing to the new ridge level. The smoky chimney was thus carried out at a proper elevation, and the whole appearance of the exterior of the house was by this means much enlarged; two of the other chimneys, after being taken down as far as was necessary, and tied together with an iron band, were arched over in the garret, and grouped above the ridge into one double stack, as shown on the upper view. The roof was projected all round, and fitted with brackets. The ventilator was placed where shown, hoods were arranged over a few of the windows, the verandas were somewhat improved, and the addition of a plant cabinet to the library completed the work as far as carpenters and masons were concerned. Both new and old parts were then painted and sanded in quiet, neutral tints, so that all appearance of alteration was at once avoided. Thus it will be perceived that, without much tearing to pieces, a new character may be given to a house, if it is only well built at first; and this design is a proof that such a result may be gained at a cost that shall not be unsatisfactory to the proprietor, for on making some inquiries for the purpose of this work from Mr. Ramsdell some years after the house was altered, I received from him a letter which touches on this point among others, and which I take the opportunity of introducing here:

"OFFICE OF THE NEW YORK AND ERIE RAILROAD COMPANY,
"NEW YORK, *September 24th*, 1855.

"DEAR SIR,—You are quite at liberty to introduce the design made for remodeling our homestead into your proposed work. I do not know how far a single instance of this sort is calculated to bear on the general subject of altering old houses, but our own experience is certainly satisfactory. We have now an agreeable arrangement of rooms, with all the minor conveniences that so materially help to make a country house enjoyable, and should be unwilling to undo any thing that has been done. The exterior of the house, which, I think, looks better than in your engraving, suits us very well, and, I dare say, may serve to show how much can be done in the way of alteration by very simple means. I can not readily give you the particulars of cost you ask for, as no contract was made, and the accounts include much other work that was being executed by our mechanics at the same time, but have no hesitation in saying that the general result is well worth the outlay incurred. The most important part that I feel to be gained is, that, with all its alterations, it is still the old homestead of the family (my father-in-law, Thomas Powell, Esq., has resided there for nearly fifty years), and we are thus enabled, without any sacrifice of enjoyment, to retain the pleasant associations that are connected with a home that has been in the family for so great a length of time.

"Very truly yours, etc.,
"HOMER RAMSDELL.

The vignette shows a method of altering a common and awkward looking form of cottage roof that is very easy of execution, and has been found, in practice, to add much to the light and shade, and general picturesque character of an old house.

STUDY FOR COTTAGE ROOF ALTERATION.

DESIGN No. 18.—(D. & V.)

PERSPECTIVE VIEW.

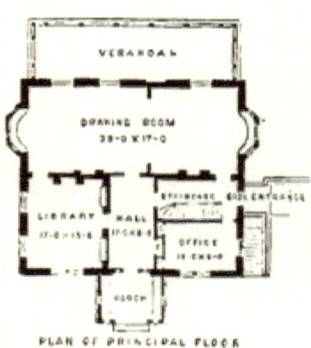

PLAN OF PRINCIPAL FLOOR.

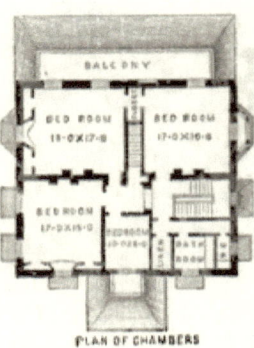

PLAN OF CHAMBERS.

DESIGN No. 18.

(D. & V.)

PICTURESQUE SQUARE HOUSE.

This house, which now belongs to a gentleman residing in Newburgh, was planned for another party in the first instance, and was partly executed with the idea that it was to be very simply and economically finished. It was commenced without any intention of constructing the dormer-windows, the projecting hoods, or the covered balcony over the lower bay, all of which, as may be seen on reference to the sketch, help materially to give individuality and completeness to the design. The main outline of the plan is a simple parallelogram, without any break in the walls, and the study may, therefore, be interesting to those who like a generally picturesque effect in a house, but who wish to avoid irregularities in the internal arrange-

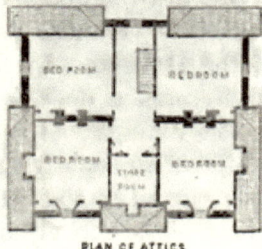

PLAN OF ATTICS.   PLAN OF ROOFS.

ment, or uneconomical projections in carrying up the brick-work. During the progress of the work the building changed hands, and came into the possession of its second owner, Mr. D. Moore, and in accordance with his instructions the design was improved in many important points. The additions already referred to were made, another bay-window was introduced, the roof to the veranda, also, was curved, and finished with a balustrade.

The plan may be thus described: An inclosed wooden porch, shown to a larger scale at page 72, leads into the principal hall, which is paved with red, black, and buff encaustic tiles in a simple but effective star pattern. This hall communicates with the library, which is a handsomely-finished room, containing two book-cases recessed in the walls on one side, and a third arranged between the windows on the other side. The architrave mouldings of the doors, windows, and book-cases being boldly relieved, and so arranged that they may group together in several different combinations. The drawing-room extends the whole length of the house. The southern part was originally proposed for a bedroom, but when the property came into Mr. Moore's possession it was connected with the parlor by an ornamental wooden arch, without folding doors. From this drawing-room the windows open on to a wide, spacious veranda, commanding an extensive view of the Hudson. In the staircase hall is a garden entrance, and a door to a small private room or office. The dining-room is in the basement, but the ground falls off so rapidly that the side of the room which looks out on to the ornamental garden, and the river beyond, is entirely out of ground, and communicates with a brick piazza supporting the veranda above. It has not, however, been thought necessary to give a separate plan of this floor, which contains kitchen and other offices, cellar, vaults, and furnace-room. Conveniently situated underneath the pavement of side-entrance is an outhouse and necessary, approached from the basement through a vine-covered veranda passage, and arranged on a similar plan to that already described in the opening chapter, and in the description of Mr. Warren's house. Mr. Moore's house was, however, the first in which this plan was tried, and it was while endeavoring to overcome the difficulties suggested by his needs that the idea occurred to me. The filling up and grading about the house was thus made more satisfactory, and the outbuildings were entirely concealed from view, which could not, in any reasonable time, have been done by trees or evergreens on account of the continuous fall in the ground, which made it necessary that the principal rooms and veranda should look down on to the lawns and garden ground surrounding the house.

The chamber plan will be found to contain four bedrooms, with a linen-room, bath-room, and water-closet, all easy of access, but planned with a special regard to privacy. An open and airy stairway to the attic leads into a roomy, well-lighted upper hall, communicating with four large bedrooms and a store-room. Above this again is a well-ventilated garret four or five feet high, that affords convenient stowage for trunks, and furnishes a complete shield from the heat and cold. The attic rooms in this house are as convenient, and almost as agreeable, as the principal bedchambers below them; and as they naturally command a more extensive view than can be obtained from the rest of the house, they have this one great advantage over the other rooms. A reference to the plan of roofs will show how, by the arrangement of the flat on the top, the whole composition is simply treated, so as to offer as little resting-place as possible for the snow where it is likely to do any harm. This house is painted in soft, quiet tints. The walls are very light buff, with a tinge of green in it, that, although scarcely noticeable, materially improves the general effect, as it makes the contrast between the house and its surroundings less decided. The woodwork is in various shades of brown. I feel, however, that it is useless for me to attempt, by words, to convey an accurate idea of any delicate effects of color that may be used with advantage in rural architecture. One might as well try to describe an Irish melody or a fantasia on the violin. All varieties of form may be explained, to a considerable extent, by simple diagrams and descriptions, but minute refinements in color refuse to be penned down, and must be seen to be understood and appreciated. This house, which was built by the day, and under the superintendence of the architect, from the time it came into Mr. Moore's possession, has been very carefully and completely fitted up throughout with all the modern improvements of gas, water, speaking-tubes, furnace, and ventilators, under the direction and instructions of the proprietor, who took a personal and constantly active interest in the work from the time he commenced his improvements till the whole was completed to his satisfaction; and it affords a fair example of what may be done in this style with a simple, straightforward plan, although it is difficult in so small a drawing to convey any very complete idea of its actual effect, which depends a good deal on the care bestowed in working out the details.

In this case the locality in which the house was built had already been provided with gas-works, and the pipes were therefore introduced as a matter of course. But in a majority of the situations in which country houses are built there are no gas-works at hand, and no likelihood of any being erected. It is, however, in any plan of moderate size, well worth while to introduce the pipes, as the expense is small, and the improvements that are constantly taking place in the machines invented for making gas on a small scale for home consumption, lead us to hope that ere long it may be both easy and economical to light a single house in this way with some simple apparatus that can be put up and readily attended to on the premises. This is, indeed, already done to a greater extent than may, perhaps, be supposed. When putting in the pipes it is, of course, necessary to prepare a notch in the beams to receive them; and unless proper care is taken this notch will, in many cases, be made by cutting an inch or more out of the very middle of the upper edge of each of the beams that occur between the side of the room and its centre (where the chandelier is suspended), although this plan is the most foolish that can be adopted, as it weakens the beam at the most important point merely to save a few feet of pipe and a little trouble on the part of the gas-fitter. The pipes should always be so planned that the notch, or recess, may be cut into the beam not more than a foot from the wall, where it will do no harm, and the connection to the centre of the room should invariably be made between the beams, and not across them.

The architect's commission for Mr. Moore's house was calculated on $12,000, which was about the cost without mantles, range, and grates. Sixty dollars was also charged for laying out the grounds, and the garden plan adopted may be gathered from the vignette below. It will be seen that the house appears to be considerably on one side of the lot. This occurred from necessity, originally, as the foundation was commenced on a piece of ground only half the size finally occupied; but the effect is, in execution, quite satisfactory, and it does not now appear at all desirable that the house should have been planned on the centre of the lot on which it at present stands, for the ornamental garden shows to good advantage, and the stretches of open lawns that are rendered possible by placing the house on one side have a much better effect than small plots intersected by gravel walks. On the grass in front of the house a pretty fountain is introduced, and the gates are made somewhat ornamental. The principal entrance gate is shown on page 286, and the side gate is sketched on one of the outside pages at the commencement of the present edition.

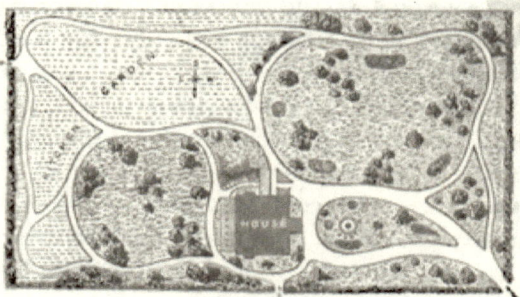

PLAN OF GROUNDS.

DESIGN No. 19.—(D. & V.)

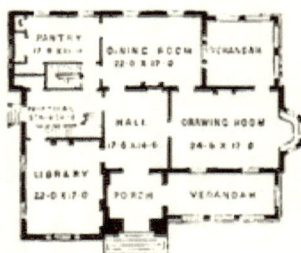

PLAN OF PRINCIPAL FLOOR

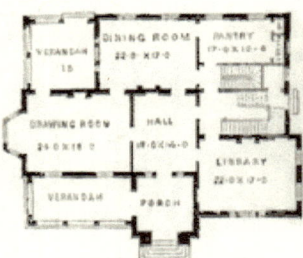

PLAN OF PRINCIPAL FLOOR

DESIGN No. 19.
(D. & V.)
SUBURBAN VILLA.

This study for a villa residence illustrates two varieties of the same leading idea of plan and elevation that were executed some years ago in Georgetown, Dist. Col. The upper one was prepared for Mr. R. P. Dodge, and the lower one for Mr. F. Dodge. These gentlemen commenced their houses about the

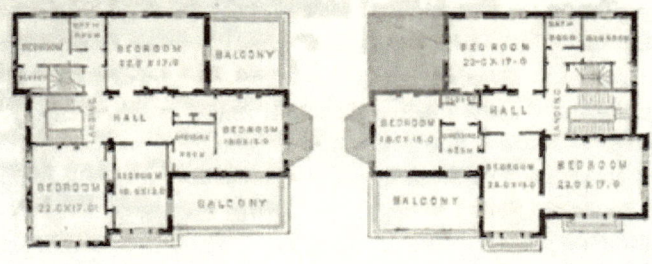

PLAN OF CHAMBERS   PLAN OF CHAMBERS

same time, and each desired to obtain the particular distribution of the rooms shown on the plan, and wished for a general similarity in the two designs, although the situations on which the buildings were to be erected differed somewhat in their local requirements. By reversing the plan, and altering the position of the library, the necessary change was made, and the details also were varied as much as possible, the windows in one design being square, and covered by projecting wooden hoods, while in the other they were made with circular heads and stone label mouldings. Minor modifications were also introduced throughout the whole of the exterior and interior; and thus, although these two houses have their principal features in common, neither is a servile imitation of the other. The plan of principal floor shows a porch that occupies the lower story of a tower, and forms a continuation to veranda on the principal front. The main hall, lighted from this porch, is of liberal dimensions, and leads to a drawing-room that is provided with windows opening on to the front veranda, and with a handsome bay at the farther end. There is also a means of access from this room to a more retired piazza, or ombra, on the other side of the house. The dining-room, which communicates with the ombra, is, as shown, entered from this parlor as well as from the outer hall, and has a large pantry, or service-room, attached. Another door leads to the staircases and garden entrance; and as the library is also furnished with a second door, all the principal rooms can be reached, as may be seen on reference to the plan, from the upper floor, without its being necessary to traverse the *principal* hall. An arrangement of this sort is calculated to add much to the privacy of the inmates of any country house, and is well worth securing when it can be obtained, as in the present case, without sacrifice of convenience in other respects. An easy, well-lighted principal staircase is provided, and a separate flight for the domestics. This latter communicates with the kitchen and offices, which are planned, according to instructions, in the basement. On the second floor will be found one large bedroom, with a dressing-room attached, three other roomy chambers, one smaller bedroom, and a bath-room, water-closet, and linen-press. Another pleasant spare room in the upper part of the tower is reached by continuing the private staircase into the attic, which affords a large open garret space, lighted from the gables. When these houses were first planned it seemed to be the intention of both proprietors to carry them out in a very simple and economical way; and as the season was a good one for building, it was roughly calculated that they might cost about $8000 or $9000. Such a plan, under ordinary circumstances, and with a perfectly plain finish, might now be fairly executed for $10,000, but not in the thoroughly complete and handsome style that was ultimately adopted by Messrs. Dodge, who were led to introduce, as the work proceeded, a good deal of cut stone into the exterior details of the design, and who appear to have spared no expense, either in labor or materials, to develop the whole idea of the plan in an entirely satisfactory manner. As the distance from Newburgh was so great, the works were not superintended by the architects, and I therefore wrote to Mr. F. Dodge, when contemplating the publication of this work, and requested him to furnish me, if possible, with some particulars as to the actual cost of these houses, and also with any suggestions for improvement that had occurred to him after a year or two's personal occupation of his own residence. In his reply he says,

"I have received your letter of the 31st ult., and will comply with pleasure with your request. The plans are with our builders. I will get them, and send by express or mail in a day or two; I mean my brother's as well as mine. We find the cost of our houses to be much beyond what Mr. Downing led us to expect—say about $15,000 each; yet we have fine houses, and very comfortable and satisfactory in every respect."

The vignette gives a slight sketch of an oak mantle-piece, introduced into a design for a dining-room, executed at Fishkill Landing. It required to be simply planned, so that it could be easily executed in the country by a clever carpenter.

DESIGN FOR OAK MANTEL-PIECE.

DESIGN No. 20.—(D. & V.)

PERSPECTIVE VIEW.

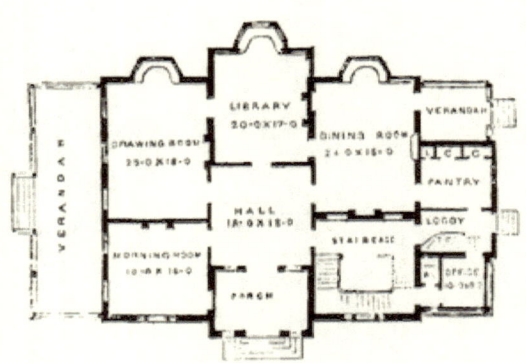

PLAN OF PRINCIPAL FLOOR.

DESIGN No. 20.
(D. & V.)
VILLA RESIDENCE WITH CURVED ROOF.

This design was prepared for an agreeable site in the vicinity of Newburgh. The accommodation proposed consists of a recessed porch opening on to a hall eighteen feet square, on one side of which is a handsome morning-room isolated from the other apartments. There is also a drawing-room, a library, and a dining-room, so arranged as to form a suite of three principal rooms, all connected together by doors, but at the same time furnished with separate entrances from the main hall. These three rooms face south, which is the pleasantest aspect in the Highlands; and as they command in this direction a remarkably fine view over the Hudson, which adds much to the value of the site, each is provided with a large bay and window seat, placed so as to overlook the river. Both the parlor and morning-room open on to a large veranda, and the dining-room is connected with a more retired piazza, as shown on the plan. A large pantry is provided close to the dining-room and servants' staircase, while a private office and water-closet are arranged near the garden entrance. An ample kitchen, with all other necessary offices attached, is provided in the basement, and a convenient chamber plan and attic are included in the design; but it has not been thought necessary to give here more than the general view, and the disposition of rooms on the first floor.

The design illustrated below has been executed near West Point, on the Hudson. In accordance with instructions, a veranda was carried all round the principal suite of rooms, and was so arranged as to connect with the entrance porch. The octagonal projection faces south, and commands an extensive view of the river.

PERSPECTIVE VIEW.

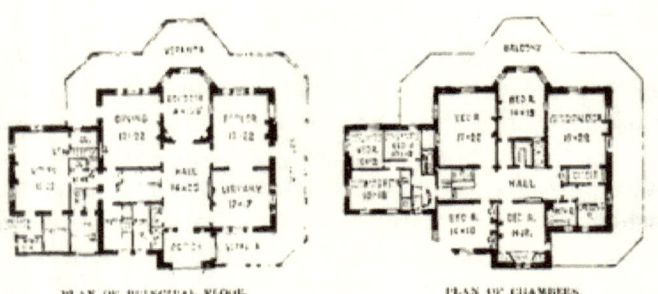

PLAN OF PRINCIPAL FLOOR.   PLAN OF CHAMBERS.

The vignette shows a sketch of a stable and coach-house, with cattle stable in the basement, proposed to be erected in Westchester County, New York. It has no very fanciful features, and is not expensively constructed, but is given here as a carefully-studied design for a simple, straightforward, roomy country barn that need not be ashamed to take its proper place in a home landscape. It is always disagreeable to see such a building, if injudiciously located, where it will be in the way of the view from the house, or have an awkward appearance from the road, or be inconvenient of access, or be so prominent that it attracts an undue share of attention. But, on the other hand, it is very agreeable to catch a view of the inferior buildings belonging to a rural home whenever they happen to be picturesquely designed, and grouped with a due regard to retirement among the trees surrounding the house. A clever and amusing chapter about "Barns" was written some little time ago for *Putnam's Magazine*, and the ideas are presented in a manner so likely to be entertaining and instructive to the readers of any work on rural architecture, that I take this opportunity to make a few extracts:

"Nothing is more essential than good impressions in childhood, and nothing secures them like a good old barn. I speak well of my mother, who was formed in a large mould, but I insist on my grandfather's barn, and I am sure that I had more pleasure in it than I have had in the new opera-house, and I would not exchange the recollections of the one for the other. My grandfather's heart was as large as his barn, and the kingdom of God was within him (I hope he is now beyond the reach of evil and selfish influence); and he knew well how necessary it was for us children to play on his hay, and he let us do it. Every Saturday afternoon my sister and I, with two other boys, played there, and on Sunday afternoon we went (she and I) to look for the eggs—for that was a work of *necessity*, and we did not then play much, for we were religious and knew the catechism. But Saturday afternoon was our high tide, and we sailed free. My sister could not climb as well as I could, but she was sagacious in discovering hens' nests, and in the art of hiding unequaled; and as she was a capital sympathizer and peace-maker, she kept her equality, and we thought her a very good fellow, if she was a girl. True, we were sorry for her, but then we said she could not help it. There was no floor but the 'threshing' floor (as in a barn built for children there should not be), but on either side of it the deep bays extended, and high up the dusky light filled the roof, through which a pencil of sunshine showed the dancing motes. In that dim space the swallows wheeled, and we watched them, hoping to scatter salt on their tails; but may be our salt was poor, may be our aim was bad, for we never caught one. We wondered what their mud nests up in the very high ridge-pole contained; sometimes young ones we knew, but eggs we always hoped, and we sighed that we could not reach them, though the old swallows took a different view of it. It was hard to climb up to the great cross-ties, and my sister could not do it; so she did not enjoy, as we did, the suspended breath of long jumps into the hay, nor the imminent peril of walking that beam. From tie to tie there were lofts, where grain was stored, made by loose poles. There was every reason to expect that we should slip through these, and fall prone twenty feet; but we never did; and this very danger gave a charm to all that was very delicious. Moreover, it stimulated our daring and educated our nerves, and was a security against the greater dangers of becoming 'spooney' good boys (not real good boys), which are apt to result in long legs, long coat-tails, long nails, and long hair, in after life—the immediate precursors of—early marriage, and other spooney good children. To be sure we always tore our clothes, and we always hurt ourselves, but we never got killed; children never do, if Providence is allowed to see to them, for the Providence of children, when they are about to fall, always tilts them into the hay, not on to the floor. Pious parents would do well to have a little more trust in Providence. Scapegraces are apt to overdo that, and to forget their own duty. My mother looked upon the injuries to our legs and pantaloons in quite a different light; and it was natural too, for the last she had to mend, and the first would 'get well.' However, we went on bravely till the shadows of evening stole upon us; then new revelations came to us, and we could not tell what large thing might not be sitting in the peak, nor what might not be lurking in the dark places, nor what those rustling noises might be, for we could hear something. Then the germ of imagination was stimulated to life, and the sublimest capacity of man—wonder—was wrought up to action; and who can tell but a poet was begun? Just at this critical moment we rushed out into the evening sky, where we found Jane milking the quiet old cow in the first light of the evening star. About my grandfather's cow there was nothing pokerish but her horns, which she shook at us now and then, so we stood by Jane and Kitty very quietly, watching the yellow milk as it streamed down into the foam (which we knew was cream); and then we walked home with Jane, not because we were afraid, but because we were good children, and wanted our suppers. Such was the end of many a delicious Saturday afternoon in that old barn, and I am old enough to love its memories. Now, in conclusion, I hope for three things. First, that when our litany is revised, immediately after famine, pestilence, and sudden death, we may be allowed to say, 'From small Gothic barns good Lord deliver us!' Secondly, that all good mothers will be sincerely sorry for what they have done, if they have supplied their boys with fringed pantaloons, a small cane, kid gloves, and long curls, instead of country air and a good barn. Thirdly, that fathers of families will read this paper, and at once begin to build roomy, ruinous old barns somewhere for their children and the swallows, and so insure good consciences, manly boys, and—my blessing."

DESIGN FOR STABLE, ETC.

DESIGN No. 21.

PERSPECTIVE VIEW.

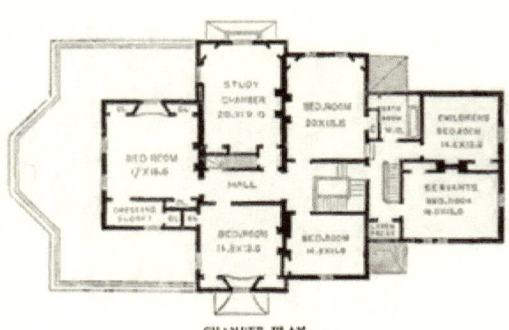

CHAMBER PLAN.

PLAN OF PRINCIPAL FLOOR.

DESIGN No. 21.
(V. & W.)

IRREGULAR WOODEN COUNTRY HOUSE.

This design was prepared for a gentleman residing at Worcester, Massachusetts, and has been executed by him, under his own immediate supervision, on a valuable site in the vicinity of that thriving city. The house was located on the edge of a beautiful pine wood that forms part of the property; and a few fine deciduous trees, that had been fostered and picturesquely grouped by the liberal hand of Nature, suggested the exact position to be selected. They now form a valuable help to the architectural composition, when viewed from a little distance, and, as anticipated, add very much to its rural, home effect. A country house built under such favorable circumstances as this may at once take its appropriate place in the landscape; and if it is agreeable in color, and designed with even an approximation to good proportion and picturesque arrangement of light and shade in its outlines, it will appear to be an old resident directly it is occupied. The leading idea of the plan was suggested by the wife of the proprietor, and the disposition of the rooms on the principal floor, with a few slight modifications, is in accordance with a pencil-sketch of arrangement furnished me, as expressing her wishes on the subject. It possesses, as will be seen, many advantages, and an explanation of it may, I hope, induce other ladies who may look over this volume to take some personal interest in the plans of the villas or cottages that are to be built for their occupation and enjoyment. The mistress of the house is, in reality, more interested than any one else in the convenience and completeness of its interior arrangement, for this part of the design is entirely under her control and regulation, and it is evident, therefore, that she must be the best judge of what will suit her individual requirements. It is well known that the daughter of Erwin Von Steinbach materially assisted her father in the design of that stupendous triumph of Gothic architecture, Strasburg Cathedral, and a sculptured memento of this interesting fact is preserved within its walls. Both father and daughter are represented in stone as consulting together on the plan of the cathedral, which Steinbach holds in his hand, together with a pair of measuring compasses, and the gently earnest and confiding expression that is to be traced on both faces is delightfully rendered. There can be no doubt but that the study of domestic architecture is well suited to a feminine taste, and it has, moreover, so many different ramifications, that it affords frequent opportunities for turning good abilities to profitable account; for if we even allow the objections that might be raised by some against the actual practice of architecture by women, such as the necessity for their climbing ladders, mingling with the mechanics and laborers during the progress of the works, and having frequently to attend to the superintendence of buildings in disagreeable weather, and at all sorts of different levels, we must, nevertheless, see at once that there is nothing in the world, except want of inclination and opportunity, to prevent many of them from being thoroughly expert in architectural drawings, or from designing excellent furniture, paper-hangings, draperies, carpets, or decorations, or from drawing or engraving on wood, or from coloring architectural perspectives in water-colors, or from modeling ornaments in clay. I do not, it will be perceived, include in the difficulties to be overcome want of natural ability, for this certainly does not exist. The tasteful and delicate needle-work that comes from the hands of women is amply sufficient proof that there is a supply of inventive capacity and artistic feeling latent among them that deserves, in civilized countries and liberally-educated communities, a much wider outlet than can be furnished by the point of a needle. To return, however, to our plan. A porch under the tower opens into a principal hall, and contains an outside door to a business office that is accessible from the staircase hall in the interior of the house. A veranda, extending round three sides of the drawing-room, also connects with this porch, so that a promenade of considerable extent is provided by the design. The hall and principal staircase are so designed that they form one symmetrical composition, and a light, open, airy effect is thus produced. The drawing, or summer room, is a handsome apartment, with windows on three sides; two of them, at opposite ends of the room, open on to a veranda, and the third is a bay, commanding an agreeable view of the distant landscape, and fitted up with a permanent settee, or lounge. The library is supplied with book-cases recessed in the wall, and has a door to the dining-room, which is supplied with a pantry and china-closet. The servants' wing is above ground, and contains kitchen, back kitchen, pantry, and store-room. In the basement are cellars and a furnace-room. The second floor plan provides in the main body of the building one bedroom, with dressing-room attached, also a large room used as a study, and three other chambers of moderate size, and all furnished with closets. In the wing will be found a bath-room and water-closet, a children's bedroom, a linen-closet, and a servants' bedroom. The attic is not finished off at present, but several agreeable rooms can be arranged here whenever they are needed.

The house was built of wood, filled in, and the interior has been fitted up carefully and completely, the work being done by the day under the proprietor's supervision. The cost has been somewhat over what was at first proposed, and an extra $5000 or $6000 could easily be spent on such a house as this by increasing the value of the external and internal work and decoration accordingly, while preparing the plans and specifications.

The vignette illustrates a termination to a gable which was used in a residence at Newburgh, described at page 225.

VERGE-BOARD.

DESIGN No. 22.—(D. & V.)

PERSPECTIVE VIEW.

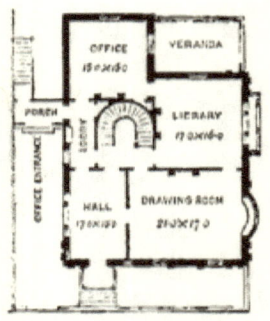

PLAN OF PRINCIPAL FLOOR.

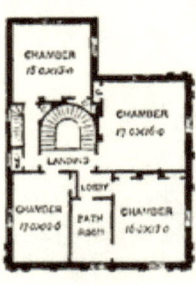

PLAN OF CHAMBER FLOOR.

DESIGN No. 22.
(D. & V.)

SUBURBAN HOUSE WITH CURVED ROOF.

This design was prepared to answer the requirements of a suburban residence for a physician, and is occupied by Dr. Culbert, of Newburgh. As it is situated on a corner lot facing the main street in the upper part of the town, it was necessary to pay as much attention to the side as to the principal elevation, and the plan has been so made that a symmetrical arrangement is arrived at on both these fronts. The house is built of brick and brown stone, the panels being recessed in the brick-work, while the pilasters and cornices are of stone. The roof is curved and covered with tin—the porch, the veranda, the railings, and the work on the roof being of iron. The plan may be thus described: A hall, 17 × 10, communicates with the parlor and staircase hall, and also with a lobby that leads directly to the consultation room. Any lady, therefore, who may inquire at the front door for the doctor, can be shown into his room at once, through this vestibule, without traversing that part of the house which is in use by the family. This lobby has also an external entrance covered by a porch, and is provided with a bell to the office, so that the greater number of those who call professionally go to and from the private room without entering the residence at all. The consultation-room communicates with the library, and this again with the inner hall, and also with the drawing-room, which is provided with a bay-window. A veranda, accessible from both rooms, is located in the angle formed by the library and office. The staircase is made circular to suit the peculiarities of plan and site, and has a good effect, as it runs continuously from the basement to the chamber floor. In Paris and other Continental cities the circular, or elliptical, staircase is in great request, and the plan offers, undoubtedly, many advantages, as it takes up less room than any other that will provide treads and risers equally easy of ascent. But I do not often introduce it in country houses, as it requires a little more care in going up and down, and stair carpeting does not adapt itself satisfactorily to the form. The extra labor needed for steps and hand-rails is also more expensive than the room and material required for other simpler styles of staircase, and no advantage is therefore gained by using it, except under peculiar circumstances. But in a design like this it may be very profitably introduced, as it enables an architect to obtain results in the internal arrangement that could not otherwise be arrived at without the occupation of valuable space. In this case, moreover, the angles supply several useful closets, which could not conveniently be dispensed with or otherwise located.

The chamber plan shows four bedrooms and a bathroom; and several fine bedrooms, protected by the curved roof, are arranged in the attic. The dining-room and kitchen offices are planned, according to instructions, in the basement.

This house has been built on the upper level of Newburgh, in Grand Street, which is at present the handsomest thoroughfare that passes through the town. It offers a drive of ample width, and thoroughly well constructed, high up above the river, and on a parallel line with it. It is lined on both sides with fine flourishing trees, most of which have been planted within the last twenty years, and many of them considerably within that time. The last four or five years' growth has wonderfully improved the appearance of this vista; and the beautiful elms and maples, with their wide-spreading and interarching branches, promise ere long to produce an effect that may equal the far-famed Hill-house Avenue in New Haven. Grand Street is thus naturally becoming the principal promenade of Newburgh, and the often-recurring glimpses of the Hudson, with its gleaming and ever-shifting freight of sails that one catches at intervals framed in the foliage of the trees on the side-streets, give it a charming pictorial character that is very rarely attainable.

This design was built some years ago, at an outlay of about $10,000; but I am unable to furnish the exact particulars of cost. This house has been painted in quiet, agreeable tints, but, in the first instance, was finished, according to the design, with a fair quality of red brick, that contrasted in color so richly and artistically with the brown stone and brown wood-work that I was very sorry indeed to see the painters at work covering it all up one fine day. There can, I know, be little doubt but that red brick, unrelieved by any other material, is altogether too vivid in color to please in an American climate; but when, as in this instance, it is used in conjunction with a good deal of stone, like the Little Falls or Connecticut brown, the effect is altogether too harmonious and satisfactory to need any attempt at improvement, and it is certain that the house will look much larger and more valuable if the work is left in its pristine state than if the surfaces are painted, although the work may be done by the best painters, and with the best colors that money can procure. Few persons take the trouble to calculate the real cost of paint, which seems a much more economical material to use than it is in reality. If the sums spent in external painting were added to the value of the brick or stone used in the building, they would often procure materials that would need no painting at all. And the same rule applies in the interior. Well-grained white pine costs as much as oiled Southern pine, and the latter is a really beautiful material when oiled or varnished, while the graining is but a sham and pretense, however well it may be executed.

The vignette illustrates one of the dormer-windows to a larger scale.

DORMER WINDOW.

DESIGN No. 23.

PERSPECTIVE VIEW.

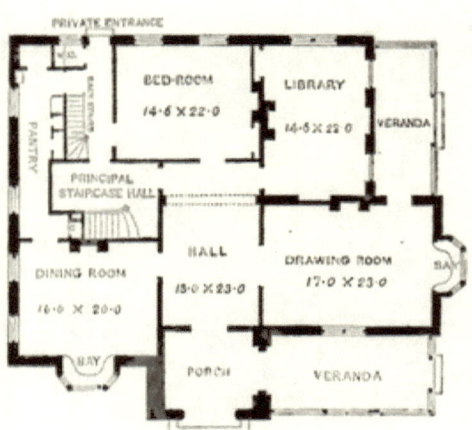

PLAN OF PRINCIPAL FLOOR.

DESIGN No. 23.

SIMPLE PICTURESQUE COUNTRY HOUSE.

This design, which is fully illustrated on a subsequent page, was carried out in a simple manner a few years ago for the residence of Mr. N. P. Willis, at Idlewild, his beautiful country place on the banks of the Hudson (with the omission of the bell turret and a few minor features that may possibly be added at some future time). The position selected, while it commands a full view of Newburgh Bay, is on the edge of a wood, and on the verge of a plateau that almost overhangs a deep ravine. The rocky sides of this gorge are clothed with sturdy pines and hemlocks, and thus often shut out from the eye, but never from the ear, a turbulent mountain stream that roars through the long, narrow pass, and falls some two hundred feet in a series of rapid, successive leaps within a short distance of the house. There is also a quieter brook that winds its way through the property, and yet another smaller rivulet, both of which, after fraternizing with their more noisy and energetic companion, flow into Moodna Creek, which almost at the same point empties itself into the Hudson. The site for the house is therefore very picturesque, and is peculiar in many respects, overlooking, as it does, the Hudson and three or four of its tributaries. These latter, with the ravine and inland scenery, lie far down on one side. Beautiful views of the bay and the Highlands are gained from the south and east fronts, and a thick pine wood encircles and protects the remainder.

When the building had been entirely erected on paper, and before the foundations were laid, all the lines of the plan were set out under the special direction of Mr. Willis, who seemed to take more interest in accommodating the house to the fancies of the genius of the place than in any other part of the arrangement. And the whole design was so fitted among the evergreens, and adapted to every peculiarity of the site, that it appears to be almost surrounded by tall, flourishing trees, although broad stretches of distance in every direction, and extensive views of the river and mountain scenery are gained from the various windows, each view being a separate picture set in a frame of unfading foliage. The advantageous result arrived at by this careful study of the exact position for such a house, although more negative than positive, is incalculable. The new house was made to look *not* new, points of view were *not* sacrificed, and time was *not* lost in waiting for young trees to grow in place of old ones that would have had to be removed for the sake of a prospect, if less foresight had been exercised at starting. In such cases it is the foot or two one way or the other that makes or mars, and when once the contractor is fairly at work, alteration is next to impossible.

Mr. Willis's house looked like an old familiar settler almost before the roof was on, and it can easily be understood that, under such circumstances, every subsequent stroke in the way of improvement will yield its fullest value.

From the upper approach road on the level of the plateau and principal entrance the house appears to be, what in actual fact it is, a plain, roomy cottage residence, comfortably sheltered among the trees; but from the lower road along the river bank, at the mouth of the gorge, its situation gives it a less ordinary effect. High up among the trees, and apparently on the very edge of a precipitous ascent, it seems to peer over the topmost branches of the dark pines, and to command the whole valley below. The position is exactly such a one as a medieval knight would have selected for his strong-hold, and a little imagination may easily transmute the simple domestic cottage into the turreted and battlemented castle. We sometimes hear a regret that the shores of the Hudson are deficient in interesting buildings, and that they lack the poetic associations that cling to the Rhine, with its thousand picturesque old ruins. This is perhaps true; but if so, it might easily be remedied, if the poetic spirit were encouraged to be active in the *life*, and not passively dependent on the *memory*, for picturesque and artistic beauty belong to whoever can realize them. It is, moreover, a little inconsistent for any true lover of freedom to take much pleasure in contemplating old castles for the sake of their associations. There are surely much more beautiful associations connected with free, peaceful industry, whenever it is generous and joyous. Dark dungeons, spiky, picturesque portcullises, and artistic machicolations for pouring down hot pitch on uninvited visitors, may, undoubtedly, have a somewhat mysterious and romantic air about them; but the sentiment they express is not, in reality, either touching or true. A nobler phase of poetic thought, and a more courteous chivalry properly belong to this freer country and more civilized era.

It would scarcely, for example, have been a very easy matter to explain to any middle age Front de Bœuf the propriety of the sentiment embodied in the following remarks of Mr. Willis when speaking incidentally of the visits of strangers to the interesting country place he has discovered and rendered enjoyable:

"To fence out a genial eye from any corner of the earth which Nature has lovingly touched with her pencil, which never repeats itself—to shut up a glen or a water-fall for one man's exclusive knowing or enjoying—to lock up trees and glades, shady paths and haunts among rivulets, would be an embezzlement by one man of God's gift to all. A capitalist might as well curtain off a star, or have the monopoly of an hour. Doors may lock, but out-doors is a freehold to feet and eyes."

The day for castles, or even magnificent mansions, has never yet dawned in America; and as its arrival must necessarily be accompanied by a return of feudalism in some form or other, any wish for the advent of such a day should be at once rejected by even the most art-loving republican. And yet, although we at once give up all hopes of this sort, no real sacrifice need be made in so doing, for a beauty of outline and color, and a picturesqueness of grouping fully equal to that which was realized by the barons of yore in their moated strong-holds, or the noblemen of the olden time in their splendid palaces, may undoubtedly be reproduced in the rural architecture needed by Americans of the nineteenth century whenever it shall be properly developed. The effect must be produced in a different way, of course, and with a different spirit to guide it, but it may, nevertheless, be equally attractive and equally poetic, if viewed from the proper point of view. In America the stranger will never, probably, be much struck with the architectural results of wealth concentrated through a series of generations; but if art flourishes as is to be hoped, he will be still more surprised and delighted by the constant recurrence of beauty and grace in the residences of the large body of the people, and the impression made on the mind will, on the whole, be more striking and more lasting than if it were excited by a comparatively few large objects of interest rearing themselves proudly up from a low general level of unprogressive poverty and wearisome monotony.

The plan provides a brick porch that may be inclosed, communicating with a veranda and with a principal hall of moderate dimensions. This hall opens into a parlor, a dining-room, a library, and a bedroom, each purposely disconnected with the other. It also leads through a door into the principal staircase hall, and a back staircase provides a private garden-entrance, and communicates with the dining-room through a pantry, in which is a rising lift connected with the kitchen. The windows that open on to the verandas are glazed to the floor, and the dining and drawing rooms have bay-windows. It was at first proposed to have a projecting window to the library, and to carry it up two stories, for this point commands a view clear down into the very heart of the glen, and in the early summer presents to the eye a wonderful waving sea of vividly green tree-tops, among which it is difficult to distinguish a single bough or trunk; but this part of the architectural arrangement had to be given up as too costly. Fortunately, however, the inexpensive green gulf, whose books are the running brooks, floats on with a glorious disregard of more conventional libraries, and never refuses to be enjoyed because it does not happen to be contemplated from an appropriate bay-window.

The dining-room was at one time designed where the bedroom is now placed, but more mature consideration located it in its present situation, so that it might receive the benefit of the afternoon sun, which, as the house is proposed to be occupied all the year round, is a desirable addition to enjoyment during the colder months in a room so much occupied by a family as the dining-room.

The library contains book-cases recessed in the walls, and, as well as the parlor, or drawing-room, opens on to a veranda more private than the one communicating with the entrance-porch. All the rooms on this floor were finished simply, the walls being prepared for papering. The casings to the doors and windows are unmoulded, and the wood-work is painted in plain tints, a liberal supply of room and plenty of fresh air being preferred throughout to any elaboration of detail.

The bedroom plan provides two chambers, with dressing-rooms attached, two other good-sized sleeping apartments, and a smaller spare room, all furnished with permanent closets. These, together with a bathroom, water-closet, and linen-room, complete the accommodation on this floor. The upper hall is open and airy, and communicates with the back staircase, which is continued to the attic story. As the roof is of a high pitch, and a large flat is designed on the top to connect the various ridges, this attic is roomy, and provides several available spare bedrooms besides what are necessary for the use of the servants; and the free, uninterrupted, uppermost hall being cheerfully lighted and well ventilated, offers an available play-room for children. This is a desideratum in a family house, as the active pattering of feet that in early years ac-

NORTH-EAST VIEW.

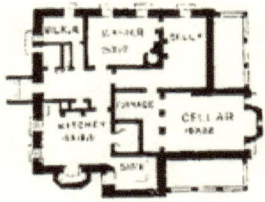

BASEMENT PLAN.

CHAMBER PLAN.

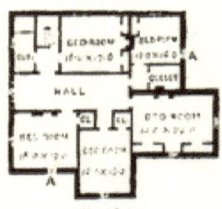

ATTIC PLAN.

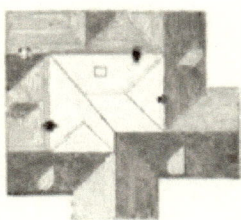

ROOF PLAN.

companies the healthy development of any new growth of possible Presidents is not less beneficial in its influence when removed a story from the ceilings of the principal living-rooms. The rain-water cistern being sunk in the floor over the bath-room, allows a large proportion of the water from the roof to be collected in it, the remainder being carried by rain-water pipes into brick cisterns sunk in the ground near the house.

The two windows shown at A A not being absolutely necessary, were omitted in execution, but they would, of course, make the two attic rooms in which they occur more agreeable, and they can be introduced at little cost, if it is ever thought worth while to insert them.

The general arrangement for the roof and the position for the proposed bell turret are shown on the roof plan. A rope could thus be arranged to pass from the bell down the side of the bath-room to the pantry below, where it would be most easily accessible when wanted.

The basement plan shows a kitchen, with windows almost entirely out of ground; a sink-room, several pantries and store-closets, a milk-room, a wash-room, a provision-cellar, a coal-cellar, and furnace-room are also provided, and an open hall communicating with an outer entrance at the lower level. This entrance is shown on the side of the plan, and was afterward covered by a porch and rendered more commodious.

The house was built of brick, covered with a lime-wash. The general plans and specifications only were provided by the architect, who was not required to furnish the detail drawings or to superintend the work. The contract for carpenter's and mason's work was taken at $7700, painting, plumbing, and other et cæteras not included.

The vignette illustrates a small cottage executed in the same neighborhood. The plan is simple, but supplies an amount of accommodation that is frequently in request. A timber porch leads to a principal hall, which communicates with a parlor and dining-room, and both these apartments open on to a veranda. The dining-room has a roomy pantry attached, that is accessible privately from the kitchen department. The chamber floor contains two large and two small bedrooms. Kitchen offices are supplied in the basement; and the contract, including painting and all items necessary to prepare the house for actual occupation, except grates and mantles, was taken at $3470.

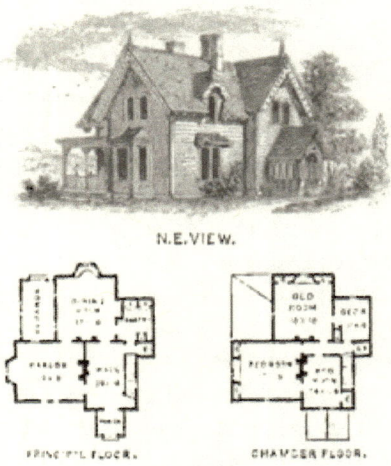

N.E. VIEW.

PRINCIPAL FLOOR.   CHAMBER FLOOR.

DESIGN No. 21.—(V. & W.)

PERSPECTIVE VIEW.

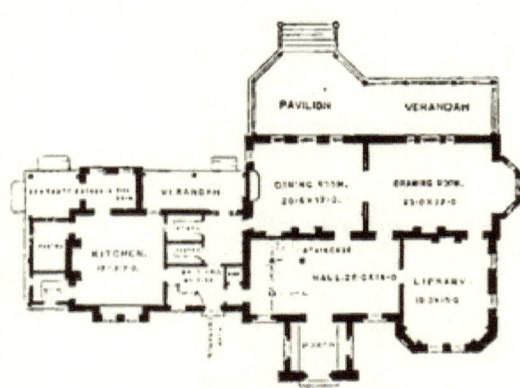

PLAN OF PRINCIPAL FLOOR.

DESIGN No. 24.
(V. & W.)
IRREGULAR BRICK VILLA.

This study has been prepared for execution for a gentleman residing in Newburgh, and is designed to suit a very agreeable site commanding an extensive view of the Hudson River. No contracts have at present been made, and I am therefore unable to give particulars of cost, but should estimate it at from $10,000 to $12,000, if the house is built of brick, and the attic left unfinished, as proposed.

The plan provides an entrance-porch opening into

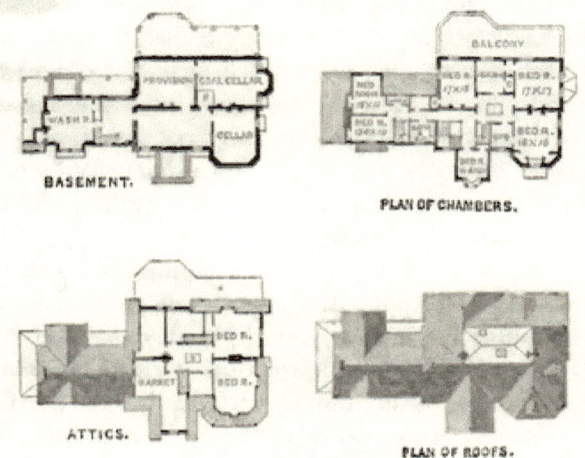

a large hall, in which is a wide, open staircase leading to the upper story. Where there is no second staircase in a country house, this plan of arranging the principal flight in the main hall is attended with several disadvantages, as the servants have to be constantly traversing it backward and forward when attending to the upper rooms. In a house like this, on the other hand, where there is a kitchen wing and a back staircase, a well-proportioned, handsome, open design, with newel posts and solid turned balusters, may be made to add much to the dignity of the hall, and to give a special character and individuality to the whole house. The owner, when giving his instructions in regard to the design now under consideration, laid much stress on this feature of the plan, and was desirous that it should be both easy of ascent and liberal in its general appearance. I have endeavored to avoid one inconvenience that is often felt in houses that have the staircase leading directly from the main hall to the chamber plan above. I refer to the want of domestic privacy that is likely to be experienced by this arrangement, if not guarded against in the plan, for the upper and lower halls being, in such cases, generally open one to the other, any conversation going on above stairs, or any slight noise that may be made is heard distinctly even at the front entrance, and this is oftentimes undesirable.

It will be observed in the chamber plan of this house that the upper hall is shut off from the staircase landing by a door, and lighted separately by a skylight. The bedrooms are thus divided off effectually from the lower rooms without losing the free, airy effect that is aimed at in the open staircase. The stairs to the attics commence from the landing, and by this means the privacy of the principal suite of bedrooms is rendered still more complete. Two bedrooms, a linen-closet, bath-room, water-closet, and house-maid's closet, with sink, are provided in the wing, and provision is made for obtaining four or five large bedrooms in the attic whenever it is thought desirable to finish them off. The roof plan is added, and it will be seen that although the house is irregular in plan there would be no practical difficulty in so arranging the lines of the roof that the snow would have no chance of lodging in any part of it. The basement plan explains itself. The three principal rooms communicate with each other, and it will be seen that the veranda which commands the river view is expanded into a semi-octagonal pavilion opposite the dining-room windows, so that the tea-table may be prepared there, when preferred, in the summer, and a small smoking-piazza is arranged so as to be easily accessible from the dining-room, which has a large pantry attached to it, connected with the kitchen, etc. The library, it will be observed, is, as it were, embayed at the end, so that book-cases may be recessed in the angles, and some variety obtained both internally and externally. The porch, with the gable over it, thus acquires the prominence that properly belongs to it. A large projecting gable over the end of the library would have had a tendency to make the entrance of secondary importance, especially as it would have occurred at the most prominent angle in driving up to the house, and the design was therefore arranged with a hipped roof, as shown on the perspective.

A projecting balcony is arranged to be entered from the room over the library. This addition to a design is not very expensive, and helps materially to give picturesque character to a front; it casts a deep shadow, and serves somewhat as a hood to the lower windows. This house, which has not yet been contracted

S

for, is proposed to be built of brick, painted in quiet, neutral tints, the cornices, verandas, etc., being of wood; and if a plain finish is used throughout, the carpenter's and mason's estimate should be about the amount mentioned above.

The vignette illustrates a design prepared for a chimney for the residence shown on page 304. As the stacks had each to contain many smoke-flues and several ventilating-flues, they would have worked out rather larger than seemed desirable for external effect if all the flues had been carried up of equal height, and the ventilating openings were therefore brought together, so as to finish on the ends of each stack at a lower level than the other flues. By this means the design of the chimney is somewhat modified from the ordinary every-day form, and a square outline avoided in a situation where it would have been inappropriate.

DESIGN FOR CHIMNEY WITH VENTILATING FLUES.

DESIGN No. 25.—(V. & W.)

PERSPECTIVE VIEW.

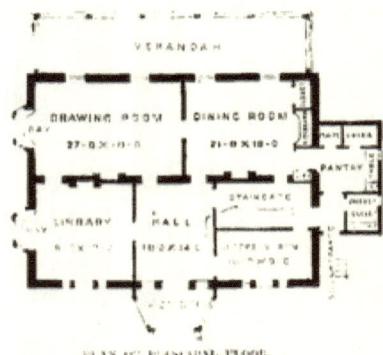

PLAN OF PRINCIPAL FLOOR.

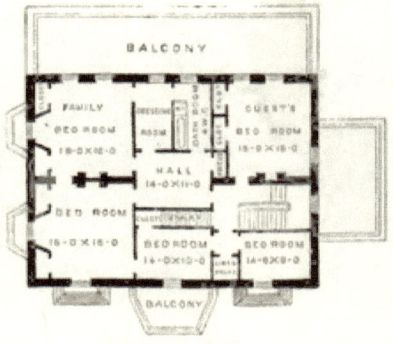

PLAN OF CHAMBERS.

VILLAS AND COTTAGES. 277

DESIGN No. 25.
(V. & W.)

SUBURBAN HOUSE WITH CURVILINEAR ROOF.

This is a preliminary study made for a gentleman in Worcester, and although it is not, and probably will not, be executed by him, the plan is one that offers an amount of accommodation that is often asked for; and as it is, moreover, a variation of the simple rectangular form, and has some peculiarities in the bedroom arrangement, it seems worth while to submit it. Almost any character of roof may be adapted to a house of this plan, and the ogee form that is illustrated in the view is effective in some situations, and has been carried into execution in a design I prepared for new roofing a square house for another party in the same neighborhood. The attic rooms in a roof of this sort are nearly as symmetrical as those in

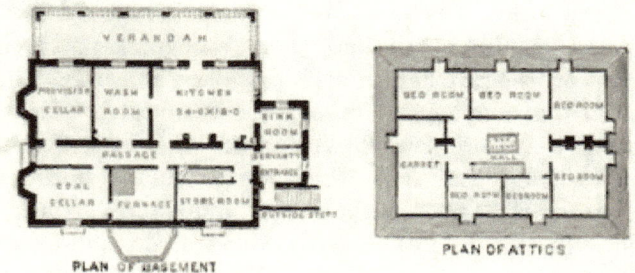

PLAN OF BASEMENT.   PLAN OF ATTICS.

the second story, and are entirely protected from the heat. The accommodation provided by the plan of principal floor consists of a projecting veranda porch, which gives access to a hall, a reception-room, library, drawing-room, and dining-room in the main body of the house, while a pantry, with dumb waiter, plate and china closets, and also a private entrance, with dressing-room and water-closet, are arranged in a one-story projection. Two of the principal rooms have bay-windows, and two communicate with a veranda. It will thus be seen that the apartments, though simply planned, are liberal in size and easy of access. The same arrangement might be adopted either on a larger or smaller scale. In the bedroom plan it was wished to provide a suite of rooms connected one with the other, and with dressing-room and bath-room attached. This, it will be perceived, is arranged for, and two spare bedrooms and a linen-press, with an easy stairway to the upper rooms, are kept distinct, with direct access from the principal flight. The attic is proposed to contain six bedrooms, a large garret, and an open hall. In the basement will be found a servants' entrance and well-lighted, airy corridor communicating with kitchen, sink-room, wash-room, storeroom, provision-cellar, coal-cellar, and furnace-room. As the study was made for a house proposed to be erected on sloping ground, there is a brick piazza in front of the kitchen, and accessible from it. This might be convenient on some occasions for drying clothes; and as it would be unseen both from the entrance front and from the interior of the house, it might be thus used without any annoyance. Such a house as this, and of the scale illustrated, would cost, under ordinary circumstances, about $10,000. It was proposed to be erected of brick, with brown stone quoins at the angles.

This design, it will be observed, is very simple in its plan, and a suburban house must generally be arranged in a compact, regular form, because the size of the lots available for this class of residence is almost always much more limited than for dwellings required to be erected in more completely rural situations. The external appearance of such a house should, I think, be somewhat symmetrical, if it is proposed to harmonize agreeably with the other buildings in its neighborhood, and with the regular and unavoidably formal line of the paved avenue, or street, that passes in front of it.

In a fine open, undulating site, well planted with trees, a symmetrical house can hardly appear to advantage, however carefully it may be designed; and in a busy thoroughfare, an irregular, picturesque plan must be equally difficult to manage satisfactorily, because the whole aspect of its surroundings will be likely to suggest the idea of precision and accuracy, and the artistic eye can hardly fail to perceive the propriety, under such circumstances, of a well-defined, self-contained expression in each of the buildings that attracts its attention.

In the suburban house an opportunity is offered to attempt a combination of both the city and the country residence; and although the plan may need to be plain and unbroken, the details may be so managed as to give any desirable degree of picturesqueness to the general composition.

The vignette illustrates a study prepared to show what may be done to give a picturesque character to an exactly square house, without any break whatever in the plan of the walls. A recessed porch leads to an open hall and stairway on the principal floor, which contains a library, drawing-room, dining-room, pantry, and small bedroom, also a water-closet near a garden entrance under the half-landing, the basement stairs being shut off at this point. The perspective view shows the veranda, or garden front. The bedroom plan gives four bedrooms, a dressing-room, bath-room, linen-press, and attic stairs. The other plans are not drawn out, as the object is sufficiently gained from the two submitted. It will thus be seen that even a simple square, which is the most absolutely formal plan that can be selected for a house, may be rendered, in a measure, picturesque, if some study is given to the arrangement of the roof lines and the spacing of the windows.

It is a common error among those who intend to build economical, straightforward houses in the country, to suppose that there is no need to consult a professional man about their plans. The fact is, that a simple design requires to be drawn out very carefully for execution, so that its internal and external proportions may be agreeable, because, in a plain house, its *proportions* are all that it has to depend on to relieve it from absolute unsightliness.

PLAN OF PRINCIPAL FLOOR.

S.W. VIEW.

PLAN OF CHAMBERS.

DESIGN No. 26.

PERSPECTIVE VIEW.

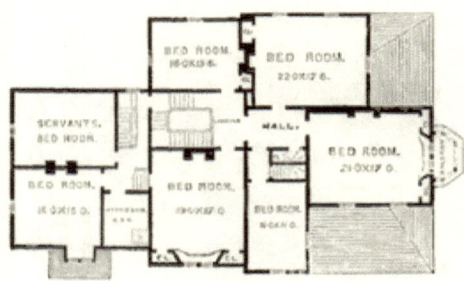

CHAMBER PLAN.

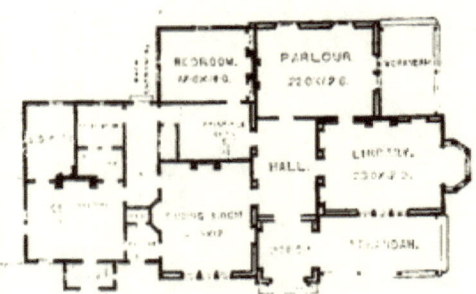

PLAN OF PRINCIPAL FLOOR.

DESIGN No. 26.

WOODEN VILLA WITH TOWER AND ATTICS.

This design belongs to a gentleman in Worcester, Massachusetts, and has been adapted to a fine situation about a mile from that thriving city. The proprietor of the site has no very immediate intention of erecting his country house, but, unlike many of his countrymen, has preferred to have his plans before him for a year or so, prior to breaking ground, so that all the minor points of internal convenience and exterior detail may be fairly and completely discussed and studied out as opportunity offers. This is a mode of proceeding much more likely to lead to a satisfactory result than to leave the whole matter to the last moment, and then hurriedly instruct an architect to prepare plans, specifications, and contracts in the course of a fortnight, that shall be warranted to supply all the accommodation required by each member of a family, in exactly the artistic form that will suit everybody, and at exactly the price that the owner has concluded it will be quite convenient for him to lay out. Such requests are not unfrequently made by employers in perfect good faith, and without any apparent perception of their impracticability, and yet it must be evident, on consideration, that if a design is wanted in a very short space of time, only a limited amount of study can properly be given to it, and the natural result is likely to be that the house will not be completely adapted to the requirements of those who propose to occupy it, simply because the architect has had no fair opportunity to consult the well-weighed instructions of his employer.

The plan shows a porch, which occupies the lowest story of the tower, and forms part of a front veranda. The principal hall connects with the principal rooms. The library and parlor communicate with each other, and with another and more private veranda. The dining-room is kept separate from the other rooms, and has an access to the kitchen wing. In the staircase hall is a door to a bedroom on this floor. The kitchen wing is arranged with pantry, store-room, etc., and servants' bedroom, that might be a wash-room, if preferred. The upper floor provides five bedrooms in the main body of the house, and a bath-room, or child's bedroom, or nursery, in the wing, with a private door opening from the family bedroom, so as to be convenient of access without traversing the hall and passage. A servants' bedroom is also arranged in the wing. A pleasant little bedroom, or study, is prepared for in the tower, and a number of rooms may be finished off in the attic, if thought advisable. The house is proposed to be built in a situation where it will have a background of fine trees, as seen from the main road; and while the home scenery is pleasantly varied, an extensive panoramic view is obtained from the upper part of the house. A tower has therefore been introduced to command this outlook; and to give a more marked character to the design, its roof has been somewhat curved, so that it may group easily with the trees in its vicinity. A two-story bay-window, it will be seen, is introduced in the design. This is illustrated in detail in the opening chapter. This study admits of many changes, without altering its main features, and it could be easily made to suit a stone or brick construction, if preferred.

To any one who is interested in the progress of rural architecture, it must be encouraging to remark how universal is the tendency now to build comfortable residences in the environs of all our large towns and cities. The love of peace and quietness, and of unaffected domestic life, that is indicated by this increasing taste for suburban houses and cottage residences, shows that it only requires a little more progress in artistic perception, and a little more appreciation of the very great advantage, to all parties, that attends study and forethought in building, to enamel the surface of this country with really beautiful rural homes. And a not distant future is, I hope, destined to show that all the liberal arts may flourish in this free republic at least as well as under the more despotic governments of the elder continent. This great step in advance might be taken in the New England States, perhaps, more easily than in any of the others; for the industry, the thrift, and the almost universal prosperity that are such leading characteristics of the people, all point distinctly to genuine refinement in private life, and to a progressive spirit of unpretending elegance that, in an enlightened Christian community, should preside habitually over every thing that appertains to the idea of "home."

The vignette shows a design for an entrance gate and piers prepared for Mr. David Moore, of Newburgh, and erected by him a few years ago. The plan of the grounds is illustrated in the vignette to Design

No. 18, and as the lot occurs at the intersection of two cross-roads, this gate was therefore placed at the angle, as shown on the garden plan, for several reasons. In the first place, it gave a more easy access to the property from every direction; and in the second, it brought the gate into such a position that a large tree on the sidewalk grouped with it agreeably, and added to the importance of the entrance. It also prevented the crowded, awkward appearance that a gate at the extreme end of one side would have had, and gave the angular view, which was, of course, the longest one, across the grounds to any one passing or entering the property. The gate is a simple design of wood and iron work, a combination which I am led to think may often be used with more advantage in rural architecture than iron alone, which, in simple, economical forms, has a very thin effect, and, when elaborated, is too suggestive of the town house to be agreeable in the country.

DESIGN FOR ENTRANCE-GATE.

DESIGN No. 27.

(SEE FRONTISPIECE.)

FAMILY COTTAGE IN THE MOUNTAINS.

This study has been prepared with reference to a particular site, and appears to be sufficiently well adapted to the present purpose of illustrating the general ideas of design that are applicable to houses or cottages built in mountainous districts. The plan is simple, and yet is not entirely symmetrical, as decided formality even in plan would be out of character with the situation. The roof, on the other hand, is both simple and symmetrical, although not formal, the break in the plan being nearly sufficient to give an impression of variety to the general effect of the whole design.

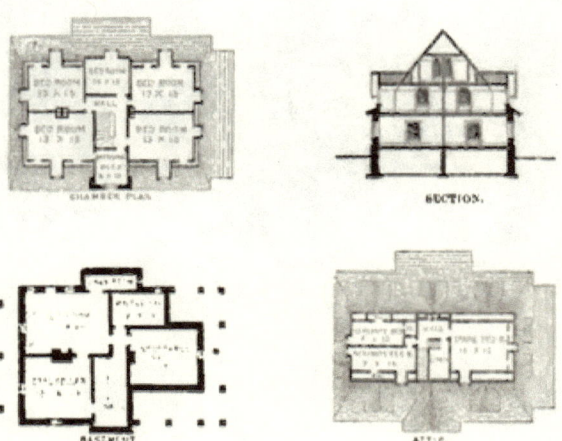

It is not desirable that the leading outlines of a house in such a position should be much broken; an impression of breadth and strength ought to be given by the general plan and by the walls, and an overshadowing, sheltering effect should be presented to the eye by the roof. Self-reliance, liberality, simplicity, and humility must be prominent characteristics in any family that spends even a few summer months successfully in a mountain home; and if such a residence is to be specially adapted to its surroundings, it must in some way or other be suggestive of these ideas.

The vignette illustrates a design for a rustic bridge that has been carried into execution in that part of the Central Park in New York which is called the Ramble.

DESIGN No. 28.—(F. C. W.)

PERSPECTIVE VIEW.

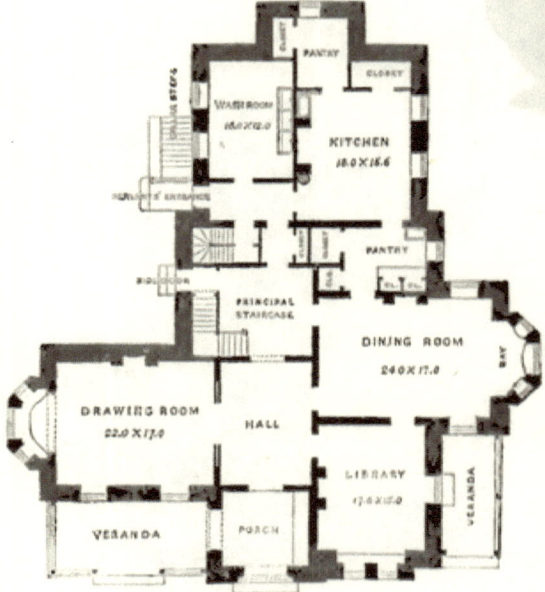

PLAN OF PRINCIPAL FLOOR.

DESIGN No. 28.
(F. C. W.)
STONE COUNTRY HOUSE WITH BRICK DRESSINGS.

This design illustrates, so far as can be done by an engraving, the effect that may be produced by the use of rough stone for the general building material, and of brick for the quoins and dressings generally.

Not having yet had an opportunity to construct a country house in this manner, although an admirer of the combination and often recommending it, I have requested my friend, Mr. Withers, to allow me to introduce into this work the accompanying picturesque example, which was built a few years ago, in accordance with his plans, for a gentleman residing at Clinton Point, on the Hudson River.

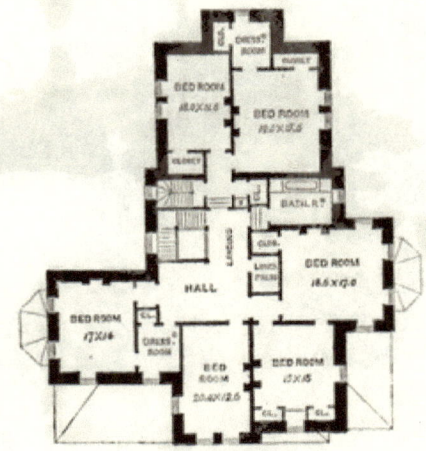

PLAN of CHAMBERS.

The vignettes show two designs for shaded seats that have been executed in the Central Park, New York.

DESIGN No. 29.

FRONT ELEVATION.

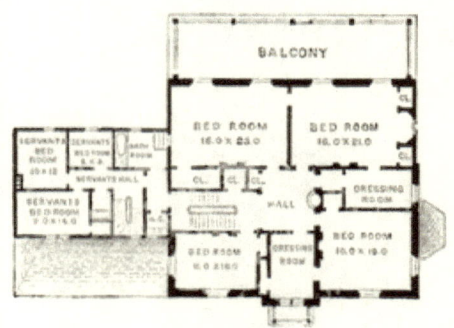

CHAMBER PLAN.

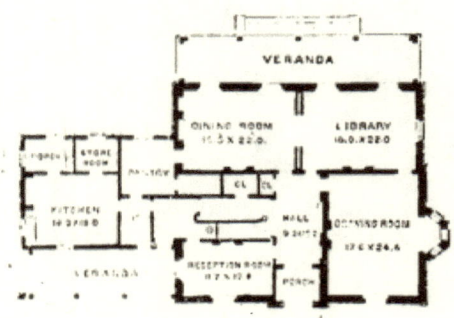

PRINCIPAL FLOOR PLAN.

DESIGN No. 29.

WOODEN VILLA WITH CURVED ROOF.

The design for this house is shown in elevation on the accompanying page, which also contains the plans of principal and bedroom floors.

In execution the effect is of course irregular and picturesque, but the geometrical drawing will show more accurately than a perspective view the true pitch that should be given to a roof of this sort.

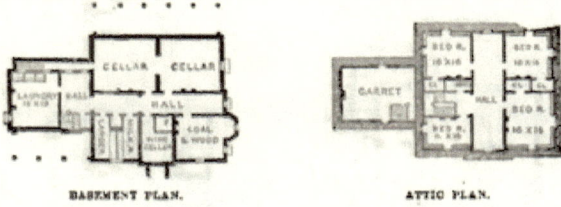

BASEMENT PLAN.   ATTIC PLAN.

The basement and attic plans are given above, and the design adopted for the carriage-house and stable is shown below.

This house was built of wood for a gentleman residing in Greenwich, Conn., and contains an amount of accommodation that is in very general request.

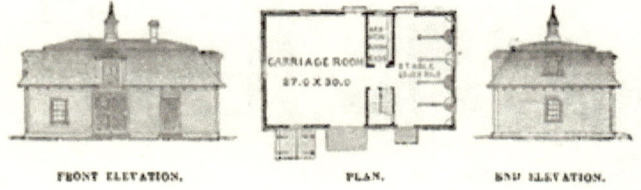

FRONT ELEVATION.   PLAN.   END ELEVATION.

The vignette illustrates a design for a square house executed at Staatsburgh, on the Hudson River. The arrangement of roof is somewhat similar to the one just described. This house contains five rooms on principal floor. The bedroom accommodation is liberal, a portion of the attic being used for guests' rooms, and the remainder, which is entirely distinct from the other part, being set apart for servants. The kitchen is in the basement.

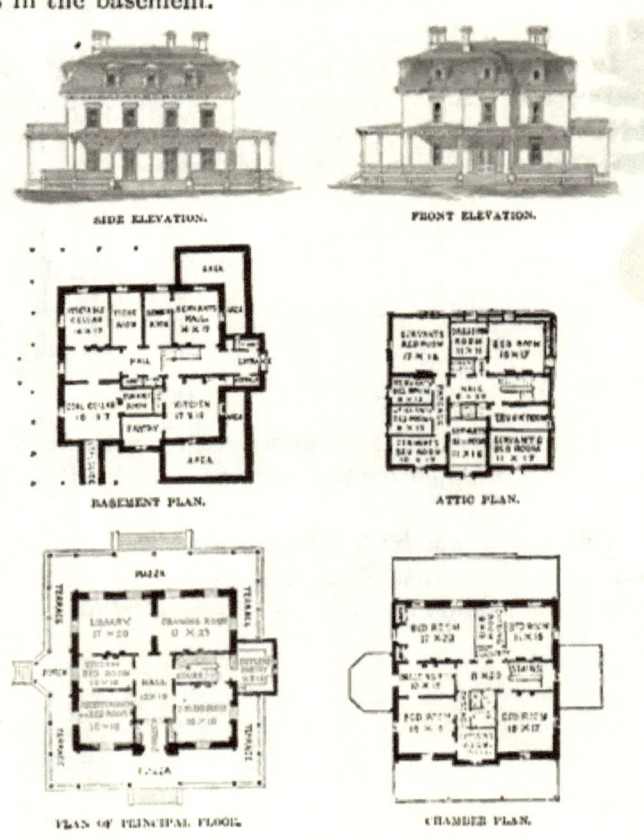

DESIGN No. 30.—(D. & V.)

PERSPECTIVE VIEW.

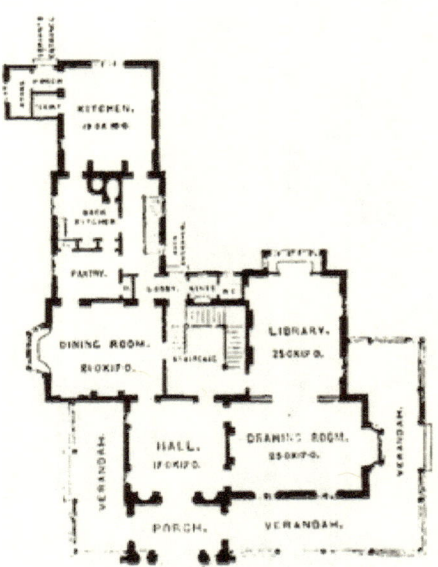

PLAN OF PRINCIPAL FLOOR.

DESIGN No. 30.
(D. & V.)

VILLA OF BRICK AND STONE.

This design was prepared for Mr. Matthew Vassar, of Poughkeepsie, and proposed to be erected at his country place, called Springside, a little to the south of the city. This estate, being full of easy sweeps and gentle undulations, is somewhat secluded and park-like in its character, fine healthy trees being scattered in groups and masses over its whole extent. These have been sparingly and judiciously thinned out by the proprietor, and the arrangement of the roads and general distribution of the grounds has been adapted to the peculiar features of the situation. The effect is very rural and homelike, although a great deal of rough work has been done, and it is only a few years since the hand of improvement was first laid upon it.

The buildings of minor importance that have been put up on various parts of the property interfere less than is often the case with the general result, each having been studied with some reference to its position and artistic importance in the landscape, as well as to its more immediately useful purpose. A roomy coach-house and stable illustrated in the last edition of Downing's cottage residences, also a cottage for a farmer and gardener, an ice-house, an aviary and poultry-yard, an entrance-lodge, summer-house and arbors, and an extensive conservatory and vinery have been erected from time to time, and the whole property has been thoroughly drained, the surface being enriched wherever it was thought necessary.

Although the property lies at some distance from the river, agreeable peeps of the gleaming Hudson and its beautiful white sails are gained here and there. Still, it is the bold horizon lines, and the broad, free stretches of richly-wooded intermediate distance contrasting, and yet in harmony, with the home landscape, that gives the peculiar charm to the place. It can, indeed, with difficulty be separated from its surroundings, and a mutual understanding advantageous to both seems to have sprung up between Springside and the scenery in its vicinity.

In country places of this size it is sometimes thought necessary to aim at increased artistic effect by a copious introduction of architectural ornaments at the salient points about the grounds; and as the result is seldom agreeable, I take the opportunity of extracting from the "Suburban Gardener" a few remarks on the subject that seem to be worthy of attention: "Architectural ornaments, such as vases, statues, etc., water in different forms, pieces of rock-work, and other objects of the like kind, form sources for varying the views from the walks of a country place; but architectural ornaments ought to be very sparingly introduced at a distance from the house, in gardens in any style, but more especially in such as are laid out in the irregular or modern manner. When mixed up with groups of flowers and shrubs, they divide the attention between the beauties of art and the beauties of nature; and as the mind can only attend to one sensation, and experience one emotion of pleasure at a time, it becomes distracted among so many. The true situation for statues is on an architectural terrace, or in an architectural flower-garden adjoining the house, the conservatory, or some other structure in which architecture and sculpture are the main features, and flowers and vegetation are altogether subordinate."

The house is approached through a porch connecting two verandas, which thus afford a lengthened covered promenade. The principal hall, seventeen feet square, with windows commanding an agreeable view, is intended to be somewhat more than a mere hall, and is designed with recesses on three sides, in which easy, simple lounges might be fitted, so that this hall, which is on the westerly side of the house, could be used as a cool morning-room in summer, if desired. The drawing-room opens from the hall, and is connected by folding doors with the library, which has a separate approach from the inner hall, and thus, when the folding doors are closed, is private and retired. The dining-room is disconnected with the other apartments, and communicates with the servants' offices through a roomy pantry. There is a private garden entrance, with a gentlemen's dressing-room and water-closet in close proximity to it. The bedroom accommodation is liberal, but it has not been thought necessary to give the other plans.

This house has been estimated, with simple interior finish, to cost about $16,000. It is designed to be built of brick, with a free use of brown stone for the angles, the copings, and the windows and other openings. The most harmonious arrangement of colors would be a soft, reddish brick, and a brown stone of as gray a tint as could be obtained. The roofs are intended to be covered with greenish-gray slates, and the eaves, veranda, and other outside wood-work should be painted of a warm oak color. There would thus be sufficient variety of color to accord with the irregular outline, and the red would have a refreshing effect in a situation secluded and sheltered among rich green trees.

The vignettes show two forms of window-hoods, the one adapted for stone construction, and the other for wood. I am not aware of any existing example in the United States of a window-hood constructed of stone; but there does not seem to be any sufficient reason why this method of obtaining a picturesque variety of light and shadow should be executed only in inferior materials, and in Design No. 33, prepared for a gentleman residing in Worcester, it is proposed to construct the window-hoods of brown stone, in accordance with the study here illustrated.

STONE HOOD.

WOODEN HOOD.

DESIGN No. 31.

PERSPECTIVE VIEW.

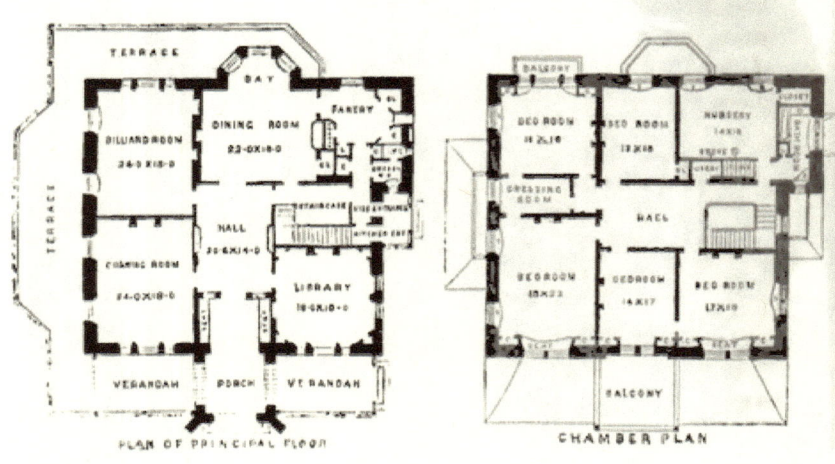

PLAN OF PRINCIPAL FLOOR.　　　CHAMBER PLAN.

DESIGN No. 31.

PICTURESQUE STONE COUNTRY HOUSE.

This design has been carried into execution for a gentleman residing at Staatsburgh, on the Hudson River. The estate is of considerable extent, the drive-road, as it passes the house, being perhaps a third of a mile from the entrance to the grounds. Still the actual building spot is somewhat limited in size, because it was necessary to select an elevated situation commanding the best views, and this happened to occur in a part of the property which was not only very varied in surface, but entirely covered by a handsome growth of trees, which it was desirable to preserve uninjured as far as was compatible with a convenient arrangement of the plan. After much examination of the different parts of the property, and due deliberation *pro* and *con*, for there were many points to be

 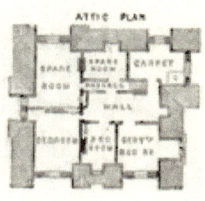

discussed, the site that seemed the most appropriate was ultimately determined on, and it then became a question how to suit the design of the house to the formation of the ground, and, so far as might be necessary, to adapt the site to the house. It seemed proper to make the plan nearly square, and without a

U

wing, for several reasons. In the first place, a sufficiently extensive arrangement for kitchen offices above ground would have rendered it necessary to cut down several more of the trees, and this, as above remarked, was to be avoided, if possible. In the second place, a wing must have blocked up the west or north views, which command the river, and are in every respect delightful; or otherwise, as it could not come on the south or entrance front, it must have been placed on the east, which is the first seen, and the most prominent at all times from the approach road. The ground was so irregular and broken, moreover, that it seemed judicious to aim at a varied outline and picturesque effect in the immediate vicinity of the house, rather than to attempt smooth extents of lawn on a level, or nearly so, with the principal floor, for this would have still farther increased the expense for filling in and grading, which must, under any circumstances, be large in such a situation. Taken altogether, it would probably be difficult to find a building spot better suited to illustrate the propriety of sometimes designing a country house with a basement kitchen. Mere economy is always in favor of this arrangement; but on level ground, unless the principal floor is stilted up some distance from the surface, the rooms in the basement will be dull, dark, and cheerless.

This house is built of blue stone taken from a quarry a few hundred yards from the building site, the stones being of various sizes and comparatively rough—the quoins, the dressings to the windows, the porch, and some few ornamental features being carefully executed in brown stone, all laid on its natural bed. The pointing mortar used in this building was specified to be of a dark red, so that by means of this warm color in the mortar joints, the cold and sombre tint of the blue stone should be modified as far as possible, and made to harmonize with the cheerful character of the rest of the house. In ten or twelve years this blue stone will begin to change its hue, and then every month will add new beauty to its color. This kind of stone is undoubtedly most harsh and monotonous in appearance when first taken from the quarry, but after about fifteen years of exposure it assumes a delicate, luminous gray tint, each stone differing just so much from the one next to it as to give life and brilliancy to the general effect in the sunlight. When this point is once arrived at, it is unrivaled as a building material, being as durable as granite, and, in connection with landscape, far more beautiful in color than any brown stone, marble, or brick.

The plan may be thus described: A porch connecting two verandas opens on to a vestibule and hall which gives access to library, drawing-room, dining-room, billiard-room, and principal staircase. The rooms are all disconnected in accordance with the instructions of the proprietor. The vestibule is fitted with permanent seats, and a terrace extends round two sides of the house. This terrace is covered by a large hood extending over it some eight or nine feet in front of one window in the billiard-room and one in the drawing-room, and reaching down to within seven feet of the floor. The shade that would be afforded by a veranda is thus obtained on this side of the house without there being any posts to interfere with the view from the windows. In the dining-room is a large bay-window recess, and the ceiling is so designed that this recess forms part of the room, and adds much to the apparent length of the house in this direction; it also increases very materially the available space for attendance on the dinner-table. The pantry is of large size, and is fitted with various conveniences, including a lift from the kitchen, several closets, and a sink.

The billiard-room is so planned that a full eighteen feet is obtained in the clear of the fire-place and walls in the narrowest part. The two side windows open to the floor, so as to afford access to the terrace, and there is a closet for cues, etc. The drawing-room is a handsome apartment, 18 × 24, opening on to the terrace and veranda. The library was originally designed to be finished with book-cases in the angles, and subsequently, at the owner's request, I furnished him with a plan for an ornamental ceiling, and a suitable design for furnishing the whole room with continuous book-cases and oak fire-place, thus carrying out the original intention in a more complete manner. There is a side entrance under the landing of principal staircase, and connected with this is a dressing-room and water-closet. A cloak-closet is also planned near here, and a lift to bring coal, etc., from the basement to this floor and the floor above. The plan is so arranged that the flue of the kitchen fire-place, which is under the dining-room, is carried away behind the closet in pantry, so that it may not heat the room unpleasantly during the summer months. A servants' staircase was thought unnecessary by the proprietor, as the principal flight is inclosed from the main rooms.

In the chamber plan will be found two large bedrooms, a dressing-room, a linen-closet, a house-maid's sink, a bath-room and water-closet, and a nursery 14 × 18. The upper hall opening on to these rooms is amply lighted, and is roomy and open, which is a great desideratum in the country, provided it can be obtained without a sacrifice of privacy.

In the attic will be found two spare bedrooms, entered near the head of the stairs, and shut off from the attic hall, which communicates with three servants' rooms and a garret.

The greater part of the basement is finished off, and supplies kitchens, sink-room, servants' bedroom, pantry, wine-cellar, coal-cellar, and furnace-room.

A contract of $13,200 was made for this house complete, ready for occupation, and with a simple, substantial finish throughout. This, however, is exclusive of the cistern and drains, the hauling and the right of quarry, all of which were furnished by the owner, and not calculated for in the estimate. All other items, such as painting, ranges and grates, furnace, etc., are included.

The vignette represents a study for a simple cottage, designed as a residence for men employed on the farm and in other operations about the estate. The cottage is in full view from the principal drive-road, and it therefore seemed worth while to consider it as an accessory in the landscape as well as a convenient home for those who were to live in it. The other side of the house showing the veranda would probably have offered a more picturesque view, but the vignette will serve to give a general idea of the simple effect aimed at. The basement, entered from a door on the outside, as shown on the sketch, was designed to be fitted up, for the use of the family, as a wash-room, and to be provided with drying-closet, ironing-room, etc. The principal floor explains itself, the two bedrooms below being for the housekeeper's use, and the chamber plan above, containing three roomy bedrooms, being set apart for the use of the men. Such a cottage, with the basement finished off, should cost about $1400 or $1500. In this instance the outlay was increased, from various causes, to $1800.

DESIGN FOR A FARM COTTAGE.

PLAN OF PRINCIPAL FLOOR.

DESIGN No. 32.

PERSPECTIVE VIEW.

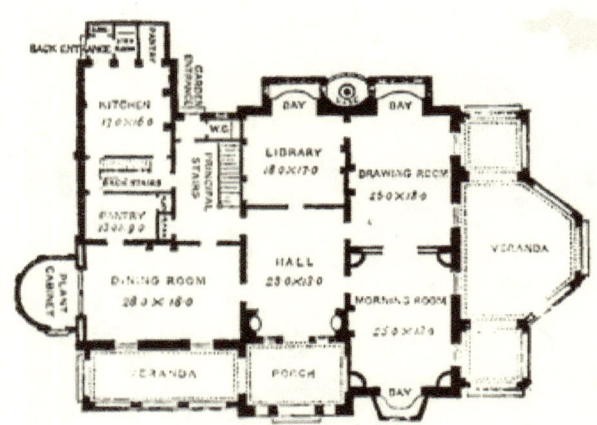

PLAN OF PRINCIPAL FLOOR.

DESIGN No. 32.

AN IRREGULAR VILLA WITHOUT WING.

This study has been prepared in detail for a gentleman residing at Millville, Mass., and is proposed to be erected in Connecticut on a beautiful suburban site, several acres in extent, on the outskirts of Middletown, which is unquestionably one of the most pleasant and attractive neighborhoods in which to build a country seat to be found in the Eastern States. Middletown is not a remarkably large place, but it possesses a cheerful and very fascinating rural character, that is to be attributed, in a great measure, to the attention that has been bestowed, from time to time, to planting in the streets and avenues. Fortunately, also, a judicious selection of specimens has been made in the first instance, which is not always the case, and

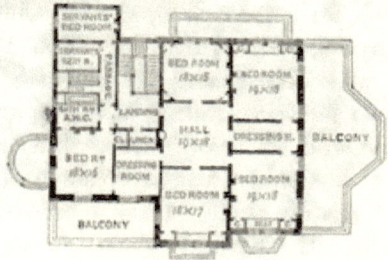

PLAN OF CHAMBERS.

the result is that the transient visitor can hardly fail to be struck with the number of fine, flourishing specimens of the different varieties of shade trees that court his attention when passing through the various avenues. It is related of Coleridge, that once, when he happened to find, in a small house in an out-of-the-way place in the country, a well-thumbed copy of one of his poems, he held it up and said, with real satisfaction, "This is fame;" and the first planters of trees on a liberal scale in Middletown might, with equal satisfaction, rejoice in the palpable evidence afforded by the present aspect of the place of the success of their influence, for they have, without doubt, been mainly instrumental in giving to the whole town that refined rural character that makes it now so desirable a place of residence.

The site on which this villa is to be erected is a choice spot in the pleasantest part of the suburbs, and consists of several acres commanding beautiful views of richly-wooded hill and dale. The drawing-room side of the house looks over a dip in the intervening ground to the Connecticut River in the distance, which will also be seen from two or three other points about the house, and will come in pleasantly in connection with the home landscape in walking about the grounds. Still, the principal river view, when the trees are grown, will always be from the drawing-room windows. The owner has interested himself for a year or two past in grading and preparing the site for the house, and, in accordance with a carefully considered plan, has laid out the roads and lawns, and planted fine young thrifty trees where they will be required to help the general effect. A wide high-road passes in front of the property, and there is a section of open ground in front of it of irregular shape, bounded by a more private road that branches out of the main avenue to several country places, and then, after making a circuit of a few hundred yards, returns again into the highway. This plot of ground will probably be planted at some time or other, by subscription, as a little park for public accommodation; and if this is done, it will add another marked feature of interest to this improving neighborhood.

The plan consists of a large open porch, that is designed in connection with the principal gable of the elevation, and is so planned that it forms a continuation to a wide veranda on the dining-room side of the house. The hall, which is arranged to have recesses for sculpture or casts on each side of the main entrance, opens on to the morning-room, the drawing-room, the library, the dining-room, and the principal staircase hall. The morning-room is a handsome apartment, eighteen feet wide, with an ornamental arrangement of the corners of the room, and a bay-window. It also communicates, through folding-doors, with the drawing-room, so that the two rooms can be thrown open together in the summer, if preferred. The drawing-room has also a large bay-window and door to the library. Both the morning-room and the drawing-room open into a large veranda, or pavilion, that is a principal feature in the design, and one, it is thought, that would add much to its desirability as a summer residence. The library is a room of moderate size, with recessed book-cases and a bay-window, designed to group in connection with the drawing-room bay. This part of the design is illustrated in detail at page 97 of the opening chapter.

The dining-room is a large room, entered from the principal hall, and also from the main staircase. A circular plant-cabinet, with an external access for the gardener, is arranged to be entered from this room through sliding glass doors, the glass being a little ornamented, but not so much so as to obscure the view of the various-colored flowers from the interior of the apartment. There is also a door to a butler's pantry that communicates with the kitchen offices, the arrangement of which will be readily understood from the plan.

In the chamber plan will be found a range of bedrooms and dressing-rooms, with bath-room and water-closet, servants' staircase, and two servants' bedrooms over the kitchen. In the attic are two or three bedrooms, but the larger part of the space is proposed to be occupied as an open garret.

This house is proposed to be built, by the day, of brick, painted, the verandas, etc., to be of wood; and it is calculated to cost about $14,000 or $15,000, finished in a simple but substantial manner.

The vignette illustrates a design for a small bath and boat house made for Mr. C. H. Rogers, and proposed to be executed by him at his country place at Ravenswood, Long Island, the lawn of which continues from the house to the water's edge, with a somewhat rapid fall as it approaches the river, so that probably a portion of the roof only of this little building would be visible from the house.

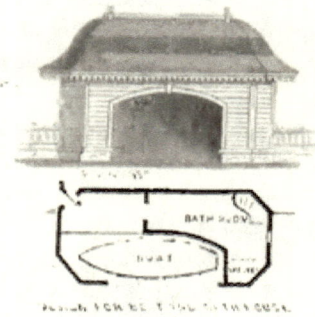

DESIGN No. 33.—(V. & W.)

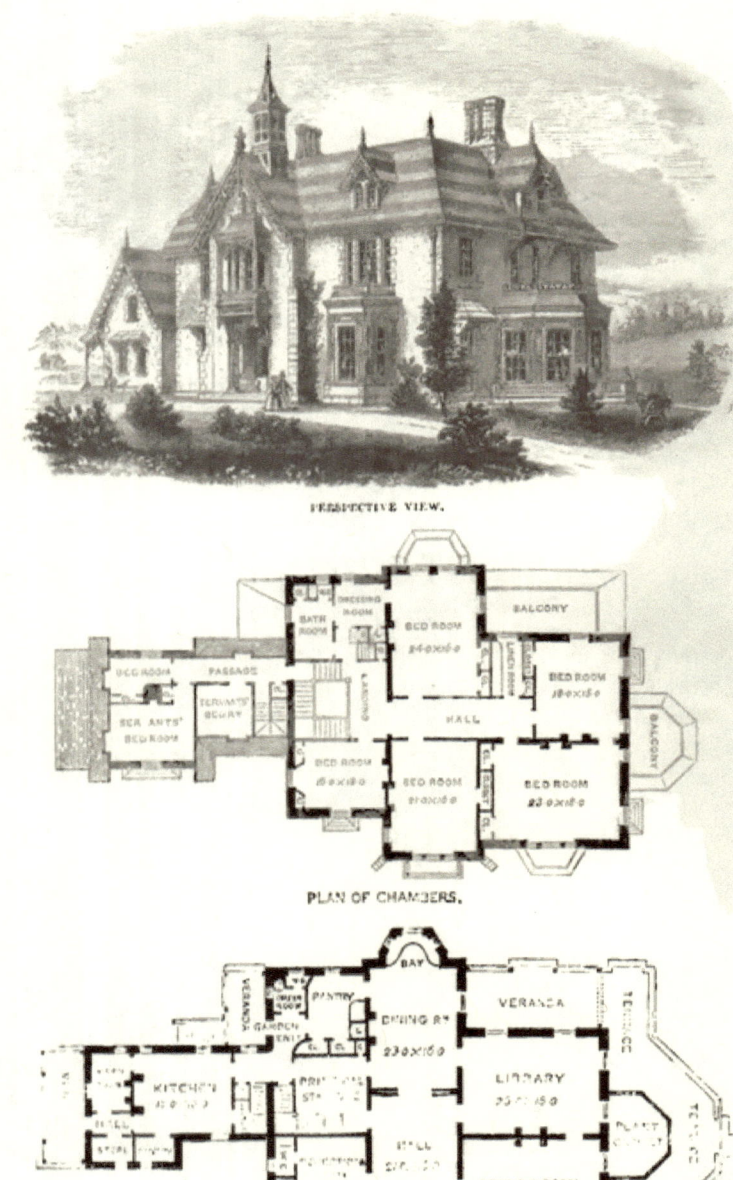

PERSPECTIVE VIEW.

PLAN OF CHAMBERS.

PLAN OF PRINCIPAL FLOOR.

DESIGN No. 33
(V. & .W)

PICTURESQUE VILLA WITH WING AND ATTICS.

This design has been fully prepared in detail for a gentleman residing in Worcester, Mass., and it is to be executed, by day's work, in brick and brown stone, with wooden trimmings to the roofs, bays, verandas, etc. The roof is intended to be covered with green and purple slates, arranged as shown on the perspective view. The plan, which is on a liberal scale, may be thus described: The principal entrance is formed by a recessed porch, which has been designed in connection with a gabled projection and a hooded balcony over the main arch, so as to give due prominence to this part of the design. The hall connects with a drawing-room, a library, a dining-room, a reception-room, and a main staircase. The principal rooms communicate with each other. The library and drawing-room have access to a plant-cabinet and terrace, and the library and dining-room open on to a veranda. The pantry arrangements are more than usually roomy and complete, as the proprietor was desirous that this part of the plan should not be in any way cramped or restricted. There is a small veranda over a garden-entrance, which has a dressing-room near it, and which can be reached from the principal staircase hall. A large kitchen and wash-room, with a store-room and pantry, complete the accommodation on this floor.

The chamber plan supplies five large bedrooms, a dressing-room, a bath-room, and a linen-closet in the main body of the house, and three inferior bedrooms are provided in the wing. The attic may contain quite a number of chambers, if required, but it is not proposed at present to finish off more than one or two bedrooms for servants. In the working plans the library is shown to be divided, by a partition, into library and study, in accordance with particular instructions, but the more simple arrangement shown here would, I think, be preferred by most persons.

The vignette illustrates the kitchen wing and the rear of the house, and will serve to show the way in which this part of the design is intended to be composed.

REAR VIEW.

DESIGN No. 34.—(V. & W.)

PERSPECTIVE VIEW.

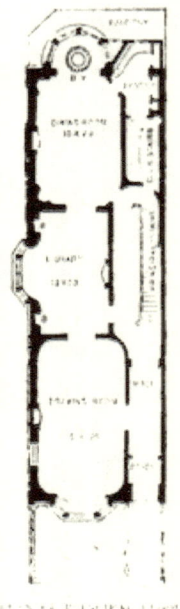

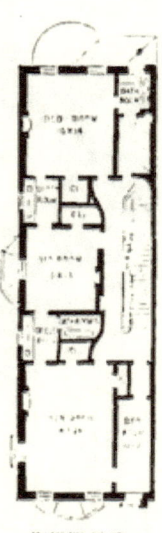

CHAMBER PLAN.

DESIGN No. 34.
(V. & W.)

A TOWN HOUSE.

This house, which is introduced as a contrast to the other designs in the book, has been executed in Fifth Avenue, New York, on a valuable lot, 25 feet wide, adjoining the grounds occupied by the Church of the Ascension. The situation for the house being thus uninclosed on three sides, it is more open and airy than is the case in the majority of house lots of ordinary width in New York, and it is, at the same time, free from the exposure and the extra expenses, such as paving, etc., that are involved in the occupation of a corner lot. It, of course, overlooks the green sward and shrubbery that have been laid out round the church, and thus, in connection with its own front garden, it may be so arranged as to avoid, in some measure, the stiffness and formality that are characteristic of most town houses, and the design may, without impropriety, aim at some picturesqueness of effect in the arrangement of the roof lines, which will come prominently into view from the other side of the street.

The house is proposed to be executed in brown stone and brick. The basement will be of brown stone, the walls of the principal floor being carried up with courses of brick and stone alternately, and the walls above principal story being of Philadelphia face-brick of superior quality, with brown stone window architraves, hoods, string courses, and chimney-caps.

The principal floor contains a vestibule, hall, staircase, drawing-room, library, and dining-room, with pantry, and back staircase to the basement and chambers.

The dining-room is proposed to be terminated by a large recess, which is to be glazed and fitted up with a stand for flowers, or a handsome vase and small fountain, if preferred. There is a peculiarity in the arrangement of this part of the design which may be mentioned here. The pantry, as well as the recess, projects from the main body of the house, for a valuable addition to the convenience of the principal floor accommodation is thus gained at a comparatively small cost; but the windows on the north, or pantry side of the semicircle would, by this arrangement, be blocked up altogether, unless some plan like the one illustrated had been adopted. This sash and the one opposite to it is glazed with ornamental glass, the two central openings being left with clear plate-glass, as there happens to be a pleasant vista view of adjoining gardens from the rear of the house. A partition is set up in the pantry at an angle of forty-five degrees opposite the sash, and a glazed opening is introduced in the outer wall of pantry in the triangular space thus shut off, as shown on the plan. The inner face of the partition is to be covered with bright tin, and the result will be that ample light will be reflected into the dining-room recess through this sash, and no disagreeable effect will be produced on the design by the convenient extension of the pantry.

The bay-window in the library, which is shown on the drawing, is not included in the working plans of the house, but can be added at any time, if approved. It would give a desirable relief to the straight, uninteresting wall adjoining the church lot, and would furnish the middle room with a little extra width and a pleasant angular view of the avenue. The principal staircase is designed with a landing in the middle of the first flight, so as to make the ascent more easy and agreeable; and a staircase shut off by a door underneath this landing leads to a fine room in the front basement that is intended to be used as a study.

The kitchen and other offices are on the basement floor, and there is a sub-cellar under the whole house.

In the chamber plan an effort has been made to get clear of lobbies and passages, and dark entrances to bedrooms; and it will be seen that by giving a curved line to each end of the partition that is planned in the central part of the house, a much larger landing than can otherwise be obtained is supplied at the head and foot of the staircase (where it is most needed), and the three principal bedrooms are entered through doors opening directly from the main hall on each floor. There are two bath-rooms on this floor, and two in the floor above, which is nearly the same in plan. The attic supplies a number of roomy chambers, and a large skylight to the main staircase. This house was carefully built throughout by day's work, and cost about $30,000 properly finished.

The vignette shows a design for a country house, for which the cellar is now being excavated, and which will probably cost about the same amount as the town house just described. It has been prepared in detail for execution by a gentleman in Orange County, New York, and is to be built on a site commanding extensive views of the Hudson. A carriage porch leads to a vestibule and octangular hall, which gives access to the drawing-room and dining-room, and to a billiard-room and library, which are separate from the other rooms. The kitchen wing is roomy, and the bedroom accommodation is on a liberal scale, as will be seen from the chamber plan. The exterior is quite simply designed, as it must necessarily be in a house of this size that is to be built for a moderate estimate. The house is proposed to be constructed with hollow walls, faced with a fair quality of brick, and relieved by a sparing introduction of brown stone where it is most required to carry out the idea of the design.

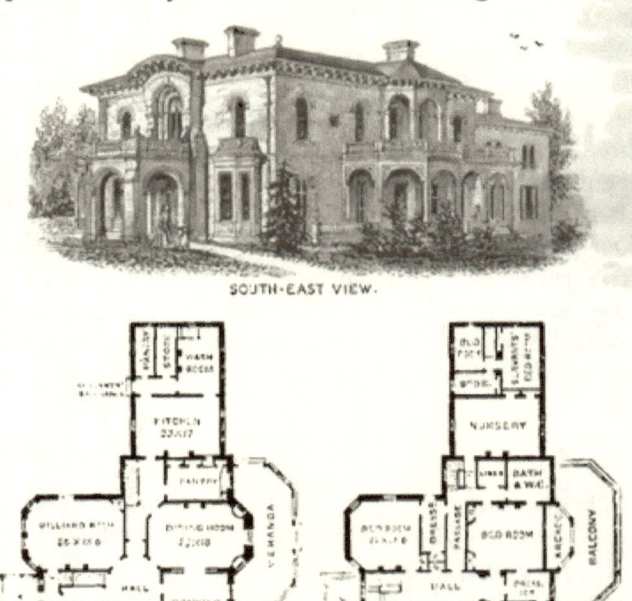

SOUTH-EAST VIEW.

PLAN OF PRINCIPAL FLOOR.   PLAN OF CHAMBER FLOOR.

DESIGN No. 35.—(D. & V.)

ENTRANCE FRONT.

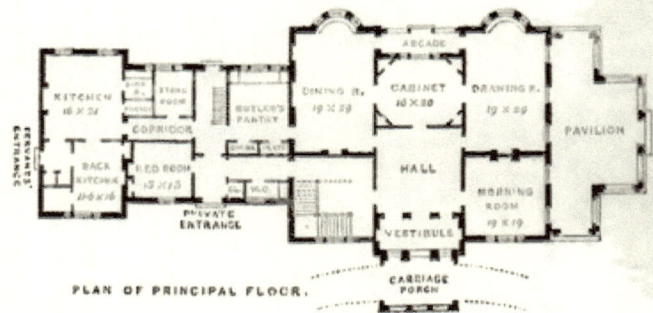

PLAN OF PRINCIPAL FLOOR.

FRONT FACING THE SEA.

DESIGN No. 35.
(D. & V.)
MARINE VILLA.

This design was erected for a gentleman residing at Newport, Rhode Island. It was built of brick and brown stone, and contracted for, including painting and plumbing, at a little under $20,000.

It was prepared for a fine situation, commanding an uninterrupted view of the sea, and including several acres of ornamental ground, that have been well laid out and planted under the superintendence of an experienced landscape gardener. A carriage-porch, of which

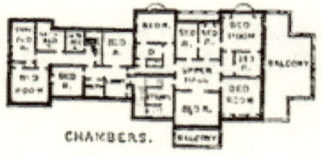
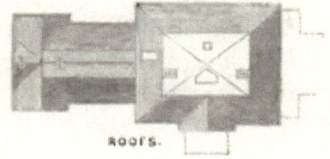

a larger drawing is given on page 74 of the opening chapter, leads to a vestibule and main hall, that contains a door to a private morning-room, and that also gives access to a suite of three other rooms of liberal size, consisting of drawing-room, cabinet, and dining-room. These apartments are connected with each other, and communicate with the lower portion of a double arcade that is introduced on the front that faces the sea. This arcade is illustrated in detail on page 99 of the opening chapter. There are bay-windows to the dining-room and drawing-room, and the latter opens on to a large veranda, or pavilion, that is one of the principal features of the design. It is thought by some that in the cool and genial climate of Newport a country house is preferable without verandas, and it is certainly desirable to admit the sunshine into all the rooms; but if a veranda is arranged, as in the present plan, so that every room that opens on to it has also another window through which the sunshine can be freely admitted, the objection to its introduction seems to be avoided, and its manifest advantages are secured without making the house dark or cheerless.

A bedroom is introduced on the ground floor of the wing near a side entrance, and the chamber plan contains a number of bedrooms and dressing-rooms, the servants' rooms being contained in the kitchen wing.

The vignette illustrates a design for the coach-house and stable, which was erected at some little distance from the house.

DESIGN FOR A COACH-HOUSE AND STABLE.

DESIGN No. 36.

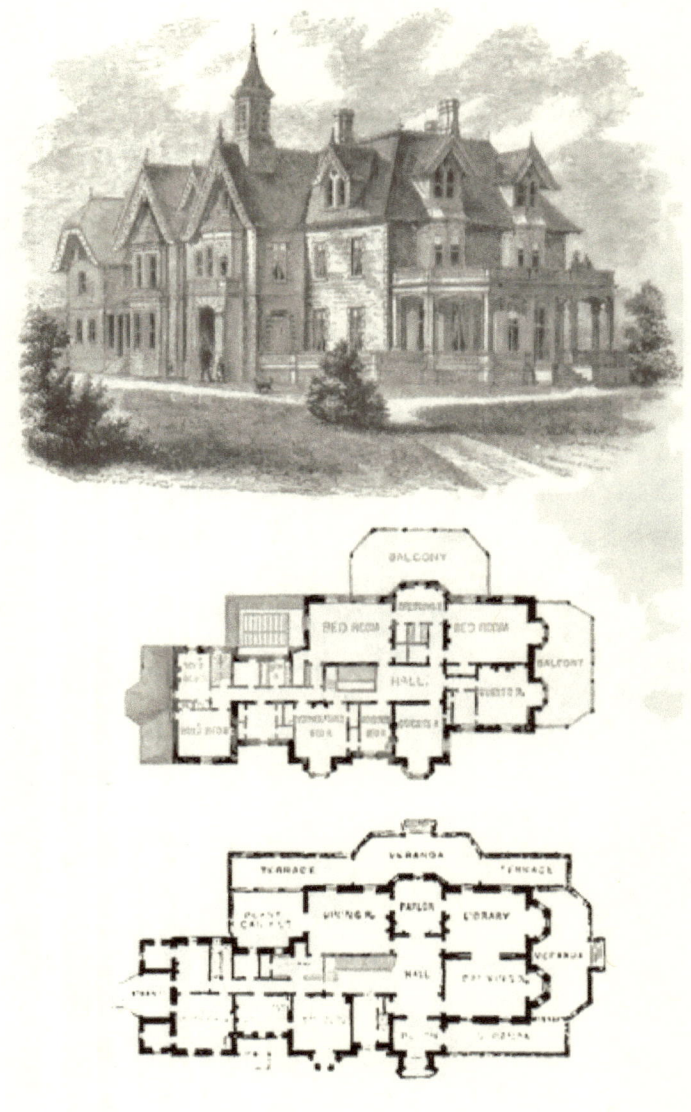

DESIGN No. 36.

IRREGULAR STONE VILLA.

This house has been erected at Fordham, near Highbridge, in the neighborhood of New York; it is constructed of stone quarried on the immediate site of the building, and the quoins and other dressings are executed in New Brunswick stone.

The building material found on the place is of a cool gray, and the New Brunswick stone is of a soft olive tint; the woodwork is painted to correspond with the latter, and as the roof is covered with Susquehanna slate, the whole effect of color is subdued, and yet pleasantly varied.

The principal floor contains four large rooms en suite, and opening on to verandas and terraces; the dining-room also connects with a plant cabinet or conservatory. A school-room and children's dining-room are provided on the other side of the house, in a convenient position with regard to the kitchen wing.

The design made for the lodge, which is also the gardener's house, is illustrated below.

PRINCIPAL FLOOR.   FRONT ELEVATION.   CHAMBER PLAN.

A design made for a coach-house and stable that has also been erected near Highbridge, but for a different owner, is illustrated below. The coach-house is 24 feet by 30 feet, and is so arranged that carriages may be driven through it when necessary. Stabling for five horses is provided, with a passage in front of the horses' heads as well as in the rear. A living room is provided for the coachman on the main floor, and two bedrooms on the floor above. An ample hay-loft is included in the design, as indicated on the plan. The materials used were rough gray stone and New Brunswick freestone.

ELEVATION.

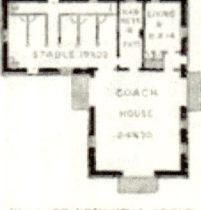 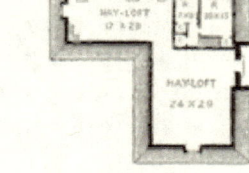

PLAN OF PRINCIPAL FLOOR.   PLAN OF LOFTS.

DESIGN No. 37.—(D. & V.)

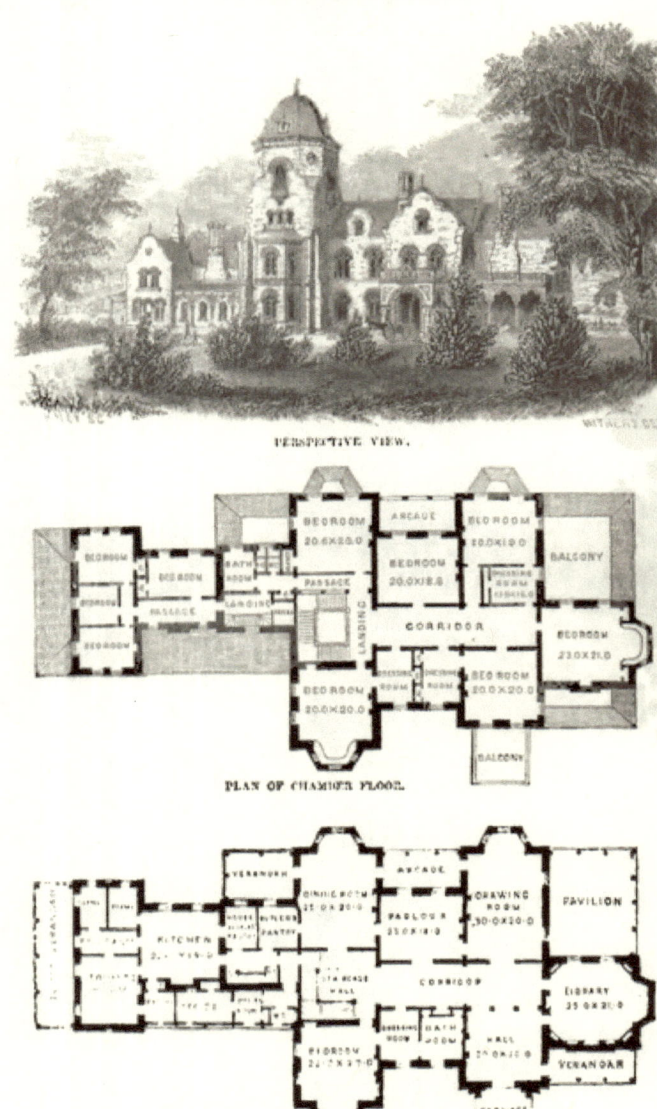

PERSPECTIVE VIEW.

PLAN OF CHAMBER FLOOR.

PLAN OF PRINCIPAL FLOOR.

DESIGN No. 37.
(D. & V.)
IRREGULAR STONE VILLA WITH TOWER.

This design, proposed to be executed on a handsome country place that overlooks the Hudson, was commenced by Mr. S. D. Dakin some years ago, but the foundations were only partially laid when the sudden death of the proprietor put a stop to the works. The house was to be built of rough stone, with cut stone dressings to the windows, and wherever else it was needed, and the whole plan was to have been carried out in a handsome and expensive style. The same general arrangement would admit of considerable variety in artistic treatment, and the exterior design might be entirely changed, without, in any way, altering the plan, which may be thus described: A carriage-porch provides a covered entrance to the main hall, which is twenty feet square. This hall is divided, architecturally, by an open, arched screen, from a wide corridor that traverses the centre of the house, and leads to the various rooms, and to the principal staircase, that is lighted from above, and is planned on a liberal scale.

The principal floor contains a library opening on to a veranda, and fitted with recessed book-cases, and it it also supplies a suite of three rooms, consisting of drawing-room, parlor, and dining-room, and a large bedroom, with dressing-room and bath-room attached. In the wing will be found a butler's pantry, a housekeeper's room, and the kitchen offices. The chamber
Y

floor contains bedrooms and dressing-rooms, the arrangement of which will be readily understood from the plan.

Such a country residence as this, economically finished inside, would cost about $30,000; and this estimate might easily be doubled in carrying out the plan of such a large house by adopting an elaborate and expensive style for the external detail, and for the internal fittings and finish of the various apartments.

The vignette shows the plan of principal floor, and the chamber plan of a square house, that would admit of varied treatment in exterior design. This study has been carefully worked out in detail, and the plans and specifications have been prepared for Mr. John W. Burt, of New York. A recessed porch opens on to a hall, 8 × 18, that contains cloak-closets, and that gives access to the parlor, the dining-room, the library, and the staircase-hall. The parlor and dining-room can be thrown into one large room at any time, as there are sliding-doors of communication, and both these rooms open on to a veranda that is octangularly arranged opposite the parlor windows, so as to give more space at this part, and to add picturesqueness to its exterior outline. The library is on the opposite side of the hall, and is disconnected with any other apartment; it is planned, however, with a door opening on to the garden-entrance lobby, which is inclosed from the hall under the first landing of the staircase; and as this lobby also contains a door to the kitchen, the library, which has a pleasant aspect for a winter room, may, if preferred, be used as a breakfast-room during the colder months.

The dining-room pantry occurs between the dining-room and the kitchen, which is in the main body of the house, but arranged, as will be seen on the plan, to be quite shut off from the living-rooms.

On the bedroom floor will be found four chambers of good size, and one smaller sleeping-room, that may be used as a dressing-room, if preferred; also a bath-room and water-closet, and an inclosed staircase to the attic, in which is provided a linen-press, a servants' bedroom, and one or two agreeable spare rooms.

The chamber windows communicate with the veranda roofs, which are nearly flat, and are furnished with a continuous railing, so as to secure an open promenade for the second floor rooms on each side of the house.

The exterior is treated picturesquely, in a simple, unpretending manner, and the house is to be built of wood, at Orange, New Jersey, on a pleasant site that forms a portion of Llewellyn Park, an estate of about five hundred acres, which has been lately taken in hand, and judiciously laid out with lawns and drives by its original proprietor, Mr. Haskell. This gentleman, after reserving in his plan about fifty or sixty acres for an ornamental park, has divided the remainder into building lots of various sizes; and as the whole area is inclosed by a boundary fence, there will be no actual necessity for separate gates to each section of the property, for the private roads that will lead from the main drive to the various country houses can be planned so as to commence in each case with a curved line running through a close plantation for a short distance, so as to insure entire seclusion, and thus the trouble of opening and shutting gates may be avoided.

I have already received instructions to prepare two

other plans for houses to be erected in Mountain Park, and this very attractive spot, which is hardly an hour's ride from the city, is undoubtedly destined to be rapidly filled up with the villa residences of gentlemen who carry on business in New York. There will be but one porter's lodge to keep up, and all annoyances will be avoided; while each resident, for a small addition to the first cost of his building lot, will hold a share legally secured to him in the public park, that is to be jointly held and superintended by the owners of the various small estates that are included within the boundaries of Mountain Park. The idea is an excellent and truly republican one in principle, and a very few years will serve to show what great advantages may result in this way from combined action in landscape gardening, if it is wisely directed in the first instance.

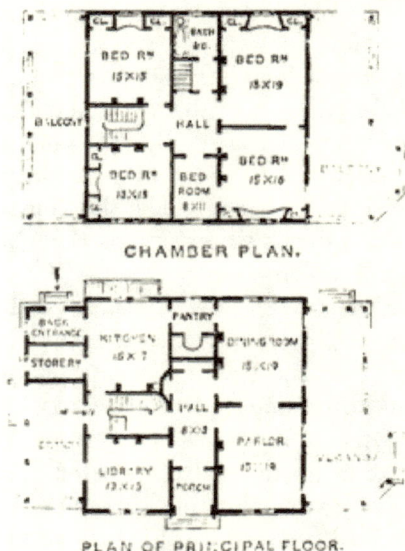

CHAMBER PLAN.

PLAN OF PRINCIPAL FLOOR.

DESIGN No. 38.

PERSPECTIVE VIEW.

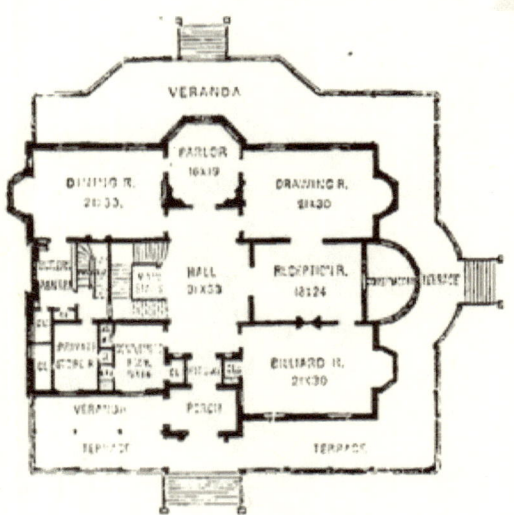

PLAN OF PRINCIPAL FLOOR.

DESIGN No. 38.

MARINE VILLA WITH TOWER.

This house has been built of brick, with Nova Scotia stone dressings, at Newport, Rhode Island, on an agreeable site overlooking the sea. The general arrangement will be readily understood from the plan on the opposite page, which provides on the principal floor a suite of five rooms, opening from a large central hall,

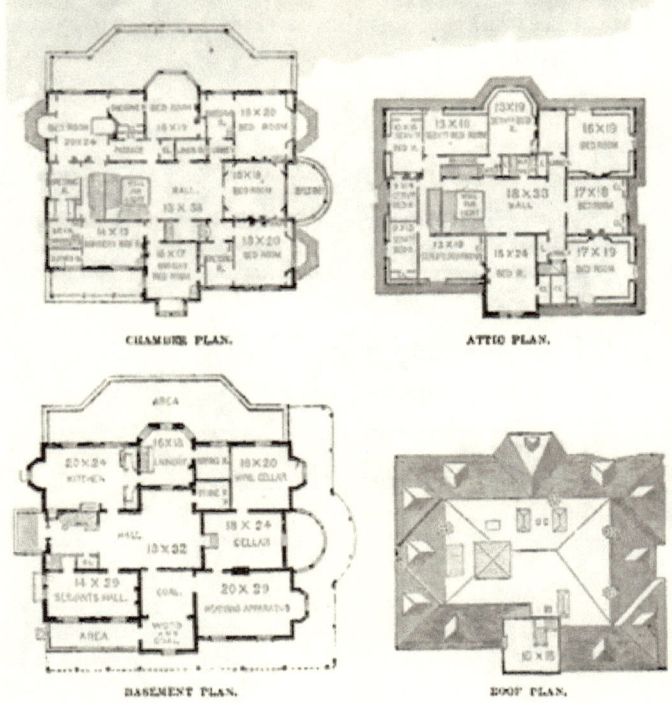

and connected with verandas and terraces, and also with a conservatory. In the chamber plan, two bedrooms, with dressing-rooms, a bathroom, and two other bedrooms, are grouped together, and separated from the rest of the accommodations on this floor; the lobbies and passages, that would otherwise be dark, being lighted from skylights above by wells marked L on attic plan. The third story contains four pleasant bedrooms and a bathroom, and six rooms (entirely separate from the main hall and guests' apartments) are arranged for servants. The kitchen accommodation is in the basement.

The plan adopted for the carriage-house and stable is illustrated below.

FRONT ELEVATION. SIDE ELEVATION.

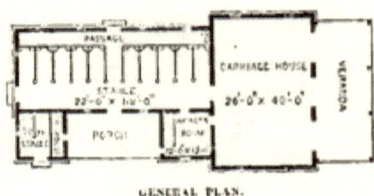

GENERAL PLAN.

DESIGN No. 39.—(D. & V.)

PERSPECTIVE VIEW.

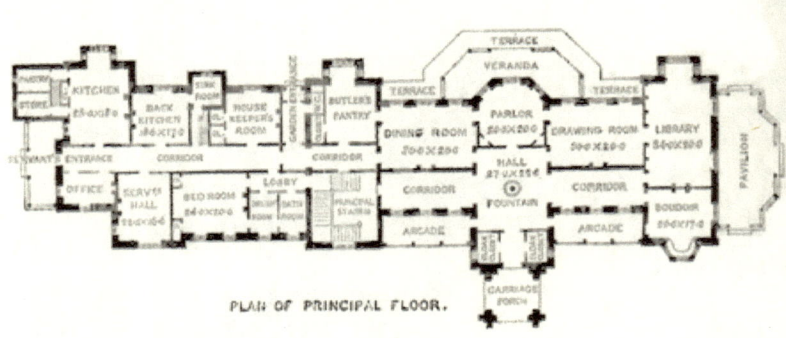

PLAN OF PRINCIPAL FLOOR.

DESIGN No. 39.
(D. & V.)

VILLA ON A LARGE SCALE.

This plan was prepared in detail some time before Mr. Downing's death, in accordance with instructions we received, but it has not been executed. It is more extensive than is usually called for in the United States, but may serve to close our examples of American villas, and to show how a large amount of accommodation may be arranged and combined. It was intended to be built of stone, on a site commanding an interesting prospect, and that sloped off considerably on one side. The garden was therefore proposed to be terraced, as shown on the perspective view, so as to do away with this sloping appearance in the immediate vicinity of the house, and to furnish a level plateau on which to build. It will be seen, on reference to the plan, that a carriage-porch gives access to an inclosed lobby, in which are cloak-closets. This inclosed lobby leads into the principal hall, which is open on both sides, to a wide corridor, and contains a fountain in the centre. There is also an open arcade, or veranda, in connection with this principal hall.

The dining-room, parlor, drawing-room, and library are designed so as to form a continuous suite of large rooms, entered from the main hall; and a boudoir, that does not connect with the other apartments, is also provided at one end of the corridor. A large pavilion is planned at the side of the boudoir and library, and a veranda and terrace, that are accessible from the three other principal rooms, are designed on the front that commands the best view. The pantry arrangements are on a liberal scale, and the kitchen offices are extensive. In the wing of principal floor will be found a large family bedroom, with dressing-room and bath-room attached, and easily accessible from the principal staircase hall and from the servants' department. It has not been thought necessary to give the other plans of this design, which was calculated to cost about $60,000 without any elaboration of finish.

The concluding vignette illustrates a design for a grave-stone erected in Newburgh Cemetery.

DESIGN FOR A GRAVE-STONE.

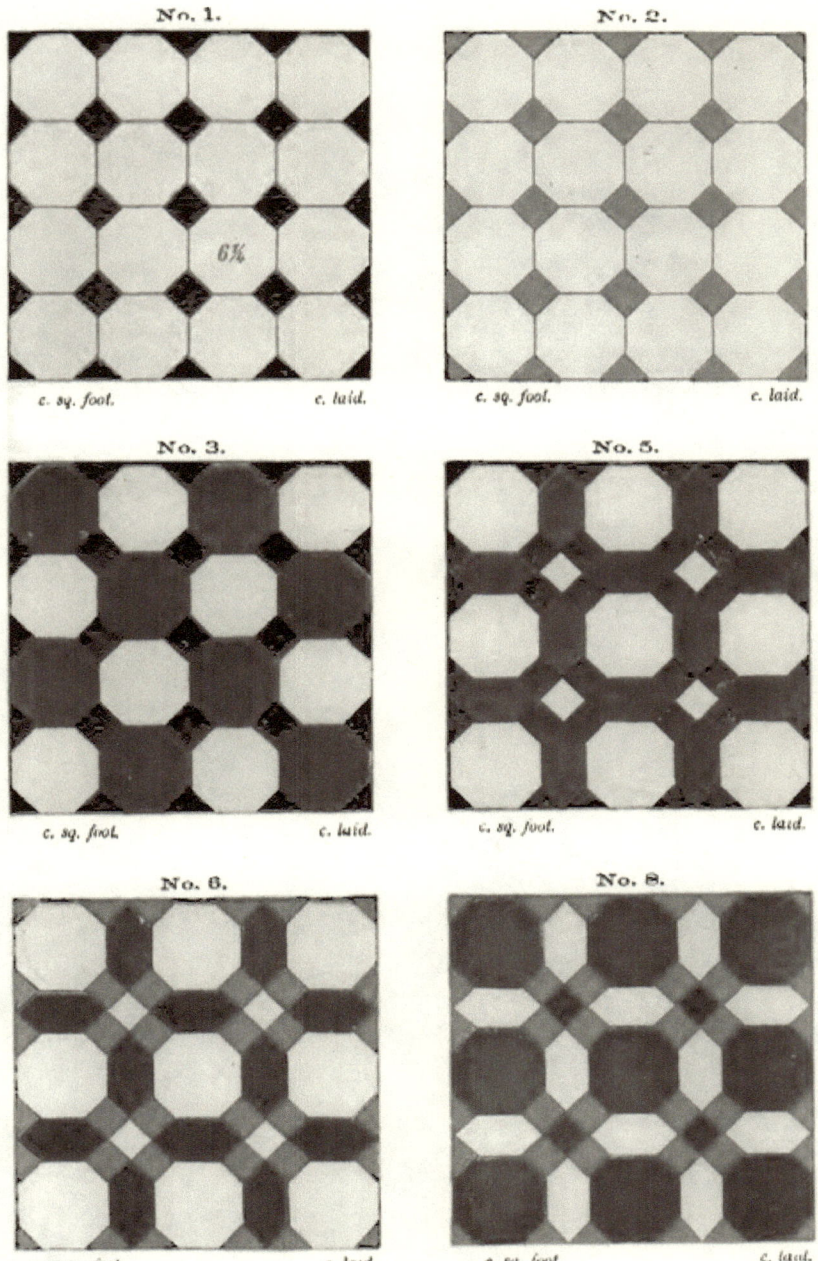

www.ingramcontent.com/pod-product-compliance
Lightning Source LLC
Chambersburg PA
CBHW021943240426
43668CB00037B/492